S0-BLN-398

FOSS Science Resources

Structures of Life

Full Option Science System
Developed at
The Lawrence Hall of Science,
University of California, Berkeley
Published and distributed by
Delta Education,
a member of the School Specialty Family

1487704
978-1-62571-318-6
Printing 3 – 5/2015
Quad/Graphics, Versailles, KY

Table of Contents

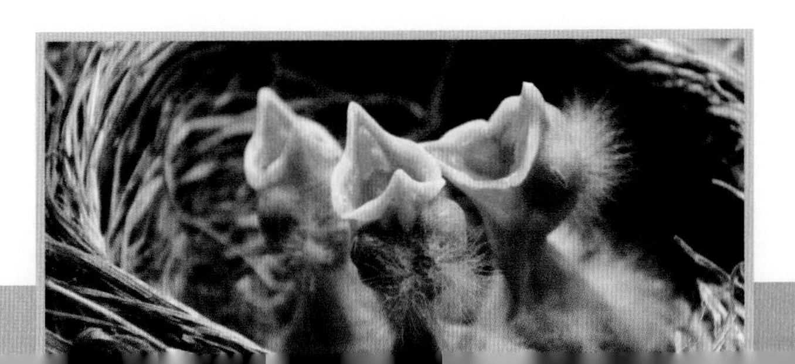

The Reason for Fruit

A fresh, sweet apricot is a treat. Peaches, plums, cherries, and apricots are favorite summer **fruits**. They are delicious and healthful. But watch out! There's a pit in the middle.

The pit of a peach or an apricot is too big and hard to eat. You have to eat around it and throw it away. The pit is an interesting part of the fruit. The pit is actually a **seed**. Do you know what is inside a seed? It's a baby plant waiting for a chance to grow.

Some fruits are not usually thought of as fruit. For instance, avocados and olives are fruits. Avocados and olives are not sweet. So why are they called fruit? Avocados and olives are fruit because they have seeds. The part of a plant that holds the seeds is the fruit. Have you seen what's inside an avocado? It has one huge seed.

Peaches

Apricots

An avocado

Olives with olive seeds

How Many Seeds?

Peaches, plums, and other pitted fruits have one seed. Other fruits have many seeds. Some grapes have three or four seeds. Apples, pears, green beans, and oranges might have six or seven seeds. That's quite a few chances for a new plant to grow.

Some fruits have dozens of seeds. Have you ever counted the seeds in a watermelon? How about in a tomato, pumpkin, or pomegranate? The kiwi fruit might have the most seeds for its size. It has hundreds of seeds.

Pomegranates

Tomatoes

Watermelons

Kiwi fruit

Green beans

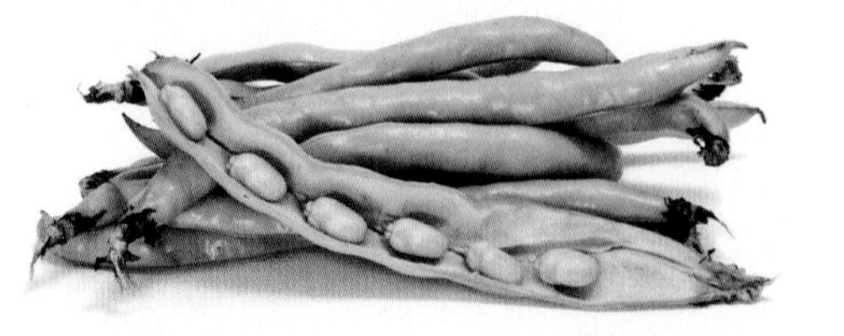

Why Do Plants Make Seeds?

Redwood trees

No plant lives forever. Some plants live for thousands of years, like giant redwood trees. Others live for only a few months, like the annual blanket flower. But each plant dies when it gets old.

Because **organisms** die, every kind of organism must **reproduce**. When plants reproduce, they make new organisms just like themselves. Peach trees make new peach trees. Tomato plants make new tomato plants. Watermelon plants make new watermelon plants. Every kind of plant makes baby plants to replace those that get old and die.

Seeds are the reproductive **structures** of most plants. Every seed contains a baby plant, called an **embryo**. The embryo in the seed is in a **dormant**, or resting, stage. You can see the embryo if you are careful. Soak a large seed in water overnight. Then carefully open the two halves of the seed. The embryo will be stuck to one side of the seed.

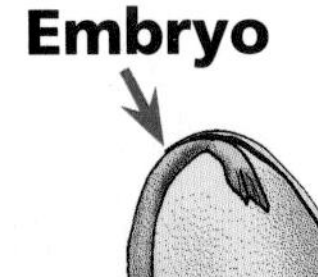

Blanket flowers

The Function of Fruit

The fruit that holds a plant's seeds is often large. The seeds in apples are much smaller than the apple. The seeds in pumpkins are smaller than the pumpkin. Fruits are also bright colors. Some cherries are red, and some grapes are purple. Why are the fruits so large and colorful?

The structure of the fruit has several **functions**. These functions help the plant **survive** and reproduce. The developing seeds need to be **protected** from weather and **predators**. Large fruits provide a protective covering that keeps the embryos in the seeds safe.

After a seed starts to grow, it needs water, light, and minerals. Sometimes the baby plant tries to grow right beneath the **parent** plant. When that happens, the baby plant has to compete with the larger parent plant. A new plant has a better chance to survive if it can move away from the parent plant. Here's where it helps to be colorful.

Cherries

Purple grapes

A pumpkin

Brightly colored, sweet fruit attracts animals. The animals carry the fruit away to eat it. But sometimes they don't eat all the seeds. They drop them far away from the parent plant. The fruit helps the plant reproduce by attracting animals to carry the seeds to new locations.

Seeds come in all sizes and shapes. Fruits come in all sizes and colors. Even though there is a great variety in seeds and fruits, their purpose is always the same. Seeds and fruits are structures that help plants survive and reproduce.

What do Thompson grapes, bananas, and navel oranges have in common? They are all seedless fruits. Sometimes an individual plant will bear fruits that don't have seeds. This is not a good thing for the plant. You probably know why. It is nice for people because seedless fruits are easier to eat. That's why seedless fruits are found in the market. Did you ever wonder how plants that don't have seeds reproduce?

Bananas

Thompson grapes

Navel oranges

Thinking about Fruit

1. What is a fruit?
2. How does a plant's fruit help it survive and reproduce?
3. What is a seed?
4. What function does a plant's seed have?

Rice plants grow in water.

The Most Important Seed

Did you know that people eat grass seeds? It's true. You probably will eat one or more kinds of grass seeds today. Wheat, corn, rice, oats, millet, and sorghum are all grasses. They are important sources of nutrition for humans. But rice is the most important. Billions of people depend on it for their food every day.

Rice was one of the first crops to be grown. In fact, it has been grown in Asia for at least 8,000 years! The countries that produce the most rice are China and India. In the United States, six states are known for growing rice. They are Arkansas, California, Louisiana, Mississippi, Missouri, and Texas.

Rice is a wetland crop. The rice plants actually grow in water. The flooded fields where the rice seeds are planted are called paddies. The rice plants are kept in the water until 2 or 3 weeks before they are ready to be harvested. It takes about 6 months for rice to grow.

This rice is ready to harvest. The rice grains are covered by a hard hull.

The rice seeds we eat grow on long, droopy **stems**. Each plant has several stems. One rice plant produces hundreds of new rice seeds. That's plenty of seeds to eat and to plant next year.

Each rice seed is covered by a hard protective shell called a hull. After the rice is harvested, the hull is removed to get to the edible grain inside. Many varieties of rice grow around the world. Some are short-grained, some are long-grained, and some are beautiful colors.

Short grain rice **Long grain rice** **A mixture of rice varieties**

Changes in the Environment

Rice is one of the most important food sources around the world. For this reason, people use a lot of land to grow rice. The **environment** changes when a rice paddy is created. **Terrestrial** (dry land) environments are changed into **aquatic** (water) environments.

Terrestrial organisms cannot live in aquatic environments. Animals, such as ground squirrels, snakes, and ants, must find new places to live and raise their young. Oak trees, sunflowers, and thistles cannot live in water. The creation of a rice paddy is **detrimental**, or harmful, to terrestrial organisms.

However, the rice paddy creates a new place for aquatic organisms to live. Crayfish and frogs live among the rice plants. Aquatic insects, such as damselflies, mayflies, and mosquitoes, **thrive**. Ducks and geese find water and food in rice paddies. Rice paddies are **beneficial** to aquatic organisms.

A rice paddy is an aquatic environment.

Frogs live in water.

Damselflies thrive in aquatic environments.

A muskrat in a rice paddy

Making a rice paddy changes the environment. Humans cause these changes. Other organisms change the environment, too. The changes to the environment can affect the well-being of other organisms.

Muskrats live in aquatic environments. They make their homes by tunneling into the banks of streams and ponds. Muskrats can live in the earthen walls that surround rice paddies. The muskrat tunnels can weaken the walls and cause them to break. When the wall breaks, the water flows out. The paddy changes back to a terrestrial environment. When this happens, the muskrat and all the other aquatic organisms must find new homes.

Muskrats causing rice paddy walls to break is one example of how an organism can change the environment. The change in this example is detrimental to the organism and to the other aquatic organisms. But the land organisms benefit because there is more terrestrial environment.

Thinking about Changing Environments

1. Grains are grass seeds used for human food. What other kinds of seeds do humans use for food?
2. How do environments change when humans make rice paddies?
3. How can muskrats change their environment, and what are some of the results?

Barbara McClintock

Did you ever believe strongly in something? Even if everyone told you your idea was silly or wrong? A scientist named Barbara McClintock (1902–1992) faced that problem for much of her life. But she never stopped believing in what she knew was true.

Barbara McClintock

Barbara McClintock was born in Hartford, Connecticut. Even when she was little, McClintock liked to do things her own way. She enjoyed all kinds of sports. Her favorite sport was playing baseball with the boys in the neighborhood. McClintock was the only girl on the boys' team. She knew that the boys didn't want her to play with them. But McClintock didn't care what other people thought. She kept on playing because she wanted to play.

McClintock did well in school, where she discovered science. When she graduated from high school, she wanted to go to college. In those days, most women did not go to college. But her father agreed that she should go. In college, McClintock studied plants and how to grow them. She loved college life. She began to focus on her studies in the field of **genetics** and graduated in 1923. She did advanced studies and received her PhD in botany in 1927.

McClintock in her cornfield

McClintock decided to become a geneticist. A geneticist is a scientist who studies how traits are passed on from one **generation** of an organism to the next. McClintock spent most of her time studying the traits of corn. She studied the color, size, and texture of corn. She grew fields of corn and studied the corn kernels (seeds). By studying the kernels, she could tell what traits were passed from one generation to the next through the corn's seeds.

In 1931, McClintock made an important discovery. Scientists already knew that every living thing passes genetic messages to its **offspring**. These messages control what the offspring look like. These messages are called **genes**. Genes are carried by structures called **chromosomes**. Scientists thought a gene located on a certain chromosome would always be there.

McClintock discovered that this was not true. Her experiments showed that genes could cross over, or move, from one chromosome to another. Crossing over meant that a greater variety of traits could exist. She published the first genetic map for corn.

Corn plants in a field

Harvested corn

In 1941, McClintock got a research position at the Cold Spring Harbor Laboratory on Long Island in New York. She worked there for the rest of her life. At the Cold Spring Harbor Laboratory, McClintock was free to do the research she loved. She often worked 80 hours a week.

King Gustav of Sweden presents the Nobel Prize to Barbara McClintock.

McClintock presented the results of her research at a meeting in 1951. Most scientists didn't understand what McClintock was talking about. Others simply didn't believe her. At first, McClintock was disappointed and surprised at the reaction she got. But she went back to her research. Once again, she didn't care what others thought. She knew she was right.

Although McClintock won several awards, her work still wasn't widely appreciated. That began to change in the 1970s. By then, scientists were able to use new technology to study McClintock's ideas. They proved what she had known to be true since 1951. It had been more than 25 years since she had first presented her ideas.

Finally, McClintock's theories were accepted by other scientists. In 1983, at the age of 81, she received the Nobel Prize in Physiology or Medicine. She was one of the first scientists to describe how genetic material controls the way an organism develops.

What Is Genetics?

Genetics is the study of how living things pass certain traits, or qualities, to their offspring. A trait that is passed down from generation to generation is called an **inherited trait**. For example, two parents with brown eyes will probably have a child with brown eyes.

Barbara McClintock continued working until her death in 1992 at the age of 90. She was always very independent and sure of herself. She spoke out about the lack of opportunities for women scientists. And she was never bitter about all the years she was ignored. "If you know you're right, you don't care," she said.

In 2005, the US Postal Service issued the American Scientists stamp series to celebrate the lives of four important scientists. McClintock was one of these scientists. A member of the US Postal Service board introduced the stamps with these words: "These are some of the greatest scientists of our time. Their pioneering discoveries still influence our lives today."

Thinking about Genetics

1. What does a geneticist study?
2. What is an inherited trait?

Nature Journal—How Seeds Travel

November 14

A curious thing is happening in my schoolyard this year. I cannot believe how many acorns there are under the oak trees. The ground is covered with them. The last few years there weren't this many. In fact, I don't really ever remember seeing this many acorns in a single year. What is going on?

I checked all the oak trees in my schoolyard and near my house. I found the same thing. I knew I needed to find out more. So I walked to my friend's house. Again, I found an unbelievable number of acorns. Then I remembered that my uncle has a lot of trees near his house. He lives about 4 hours away from me, so we e-mailed each other. Here's his reply.

Hi,

It's great to hear from you. I've got to admit, I wasn't really paying attention to the acorns this year. The squirrels and blue jays are busy eating, burying, and collecting them, and there aren't many on the ground. So I'd say we have a normal number of acorns this year. But 3 years ago my oak trees created an overwhelming number of acorns. This was true for all of the trees in my area of the state. Let me know what you figure out, my little nature detective!

Love,

Uncle Jim

Why are all the oak trees around here producing so many acorns?

I asked my teacher, and she suggested I go to the library. The librarian was almost as excited as I was. She actually said to me, "The squirrels don't even know what to do with all the acorns in my yard this year!" After a short search, we found a book about trees. We reviewed the index and went to a page about oak trees. This is what we found out. Some oak trees produce acorns every other year instead of every year. Other oak trees produce very large crops of acorns every 4 to 10 years. These same trees produce smaller crops of acorns in other years.

The book went on to say that in years when the trees produce smaller crops, the trees might have damage from insects or bad weather. In those years, the squirrels and other animals are able to eat most of the seeds. When the trees produce lots and lots of seeds, it is called a **mast year**. During a mast year, they all produce a greater number of seeds. This gives the oak trees a better chance to reproduce. During mast years, the animals that eat and store the seeds for winter can't collect all the seeds. They leave many seeds to grow into trees.

Now that the acorn mystery is solved, I've started looking around a little more carefully at how many seeds plants create. Seeds are everywhere! The maple trees have seeds that twirl away from the adult plant. A strong breeze can send hundreds, maybe thousands, of dried twirlers out away from the parent plant.

A dandelion puff ball has about 50 parachuting seeds. My brother and I once had a contest to see whose dandelion seeds stayed in the air longer. I won! One of mine traveled up into the air and out of our sight. My mom wasn't too happy with this game. She said, "Stop! You're blowing the seeds of weeds everywhere." I guess she didn't want weeds all over our yard. I think she forgot that the wind could do the same thing.

The chain-link fence at the far end of the schoolyard is covered in berries that birds love to eat. I've heard that birds will digest the fruit of the berry. Then the seed will pass in their droppings and might produce a new plant if it lands on warm, moist soil.

It's no wonder so many weeds grow in our schoolyard garden beds. Seeds have so many different ways to travel, are so plentiful, and are everywhere. That's what I've discovered about seeds in my schoolyard. What can you discover in yours?

Thinking about How Seeds Travel

Look at these pictures. How do you think these seeds travel away from their parent plants?

Germination

A seed is a living organism. To be more exact, a seed contains a living organism. The tiny structure inside a seed is the embryo. An embryo is a living baby plant. The embryo is in a dormant, or inactive, stage. The embryo is waiting for the right conditions to start growing.

But there is more to a seed. The largest part of the seed is food storage. The storage structures are called **cotyledons**. Some seeds, such as beans, peas, and sunflowers, have two cotyledons. Seeds of grasses, such as corn, rice, and wheat, have one cotyledon.

The embryo and cotyledons are wrapped in a tough outer layer called the seed coat. Some seed coats are thin, like the coat on bean and pea seeds. Other seed coats are tough and woody, like the shell on a peanut or sunflower seed. And there are seeds with coats so hard you need a tool to open them. Coconuts, walnuts, almonds, and other nuts have very hard seed coats. The seed coat protects the embryo.

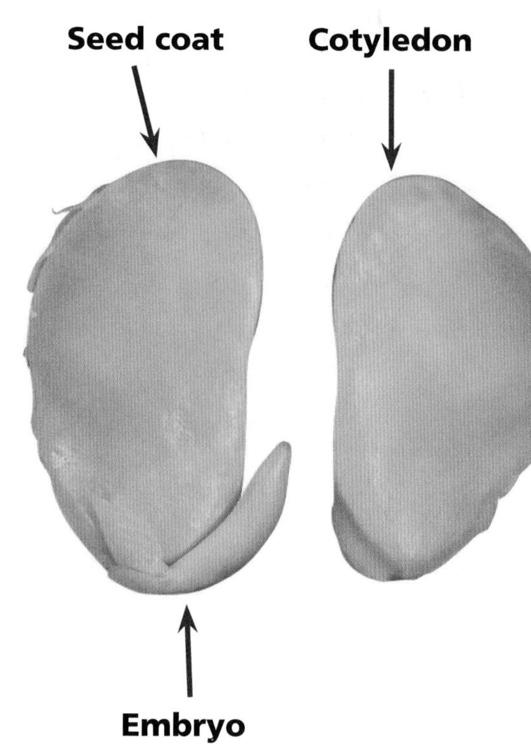

A coconut

A bowl of mixed nuts

Sunflower seeds

Starting to Grow

Germinated seeds

The signal for the embryo to start growing is water. When water makes it through the seed coat, the embryo and the cotyledons soak it up. The cotyledons swell. The embryo starts to grow. The swelling cotyledons break open the seed coat so that more water can get in. This is called germination. It is the first step in seed growth.

Soon the embryo starts to develop structures. The first structure to come out of the seed is the **root**. The root usually comes first because the root brings water and **nutrients** into the plant. Then the stem and first **leaves** come out. The stem gives the plant support as it grows.

The baby plant is now a seedling. The cotyledons stay attached to the seedling. The seedling is using the food stored in the cotyledons to grow. As soon as the seedling gets its first leaves, it can make its own food. By that time the seedling has used most of the food in the cotyledons.

The germination process is the same for most seeds. The thing that changes from seed to seed is how long it takes for the seed to germinate. Some seeds can germinate right away. Other seeds take years to germinate. Why the difference?

Bean seedlings

Germination and Environment

Pine seedlings in a burn area

Plants get only one chance to find a place to live. The place where a seed germinates is where it will spend its whole life. That's because plants can't move.

One way plants improve their chances of survival is to germinate when conditions are good. If a seed ends up in a location with good conditions, it will germinate, grow, and survive. If it falls in a poor location with bad conditions, the seed may fail to germinate. If it does germinate, the seedling might die later.

Fruit attracts birds. Birds eat the fruit and move the seeds to new locations. Moving away from the parent plant is one way to improve chances for survival.

Some seeds will germinate only after they have been tumbled and scraped over rocks. This weakens the seed coat and allows water in. Other seeds, like strawberry seeds, are weakened when they pass through the digestive system of birds. Pine tree seeds germinate in large numbers after they are heated by a forest fire. The fire kills the **mature** trees. As a result, the seedlings get plenty of light and nutrients.

Grass sprouting under snow

Some seeds come from plants that live in areas with cold winters. Many of these seeds will not germinate until they have been cold for a long time. These seeds don't germinate until spring, when the growing conditions are good.

One more thing you might know about germinated seeds is that some are excellent food. The bean sprouts sold at many markets are germinated mung beans. The sprouts are delicious in soups and stir-fried dishes.

Mung beans **Mung bean sprouts**

Mung bean sprouts ready to eat

Thinking about Germination

1. How can you tell when a seed has germinated?
2. What needs to happen to seeds before they can begin to germinate?
3. What role does the environment play in seed germination?

Life Cycles

The word *cycle* means "go around." A wheel goes around. You can observe a wheel go through one cycle. Put a mark on a wheel. That's the beginning point. Turn the wheel and watch the mark go around. When the mark comes back to the beginning point, the wheel has completed one cycle. Another cycle is the one that happens every day, from sunrise one day to sunrise the next day. One year is a cycle. The Moon goes through a cycle of phases each month.

Organisms go through **life cycles**. But an organism's life cycle is a little different from going around in a circle. Like all cycles, a life cycle has a beginning, things happen, and then you find yourself back at the beginning again.

You studied the life cycle of a bean plant. The life cycle started with a bean seed. Inside the bean seed was the dormant embryo of a bean plant. When the bean seed soaked up water, the seed germinated. The bean plant started growing.

The root was the first structure to appear. Soon after that, the first leaves appeared on the end of a stem. The baby bean plant had developed into a bean seedling. It took several weeks for the bean plant to get bigger and grow more leaves and stems.

When the bean plant was mature, it developed **flowers**. The flowers changed into fruits, called green beans. Seeds developed inside the fruits. When the fruits were mature, there was a crop of new bean seeds. The bean plant had gone through its life cycle. The plant started as a seed and completed the cycle when it produced new seeds. The seeds might grow into new plants. The life cycle repeats over and over again.

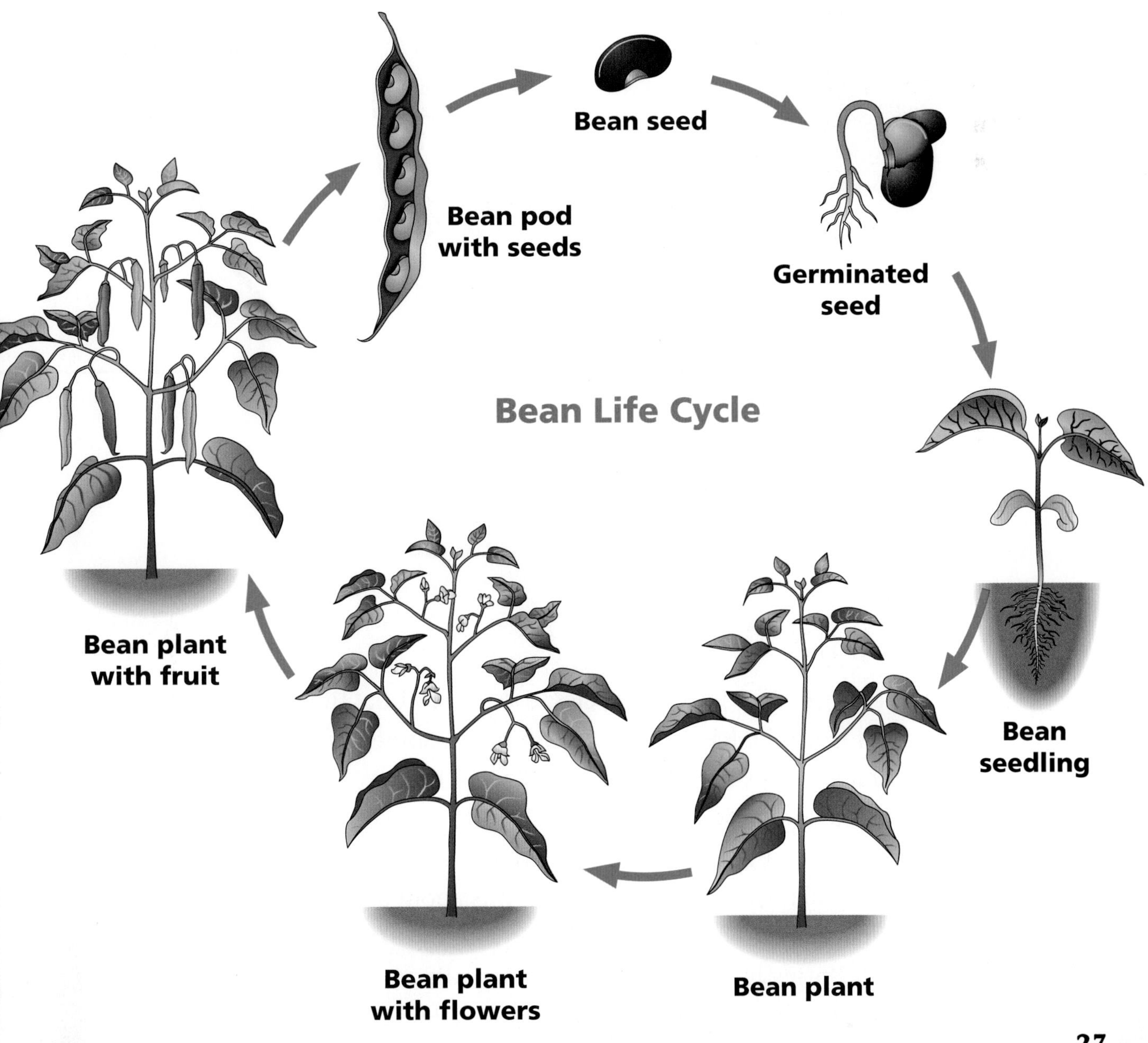

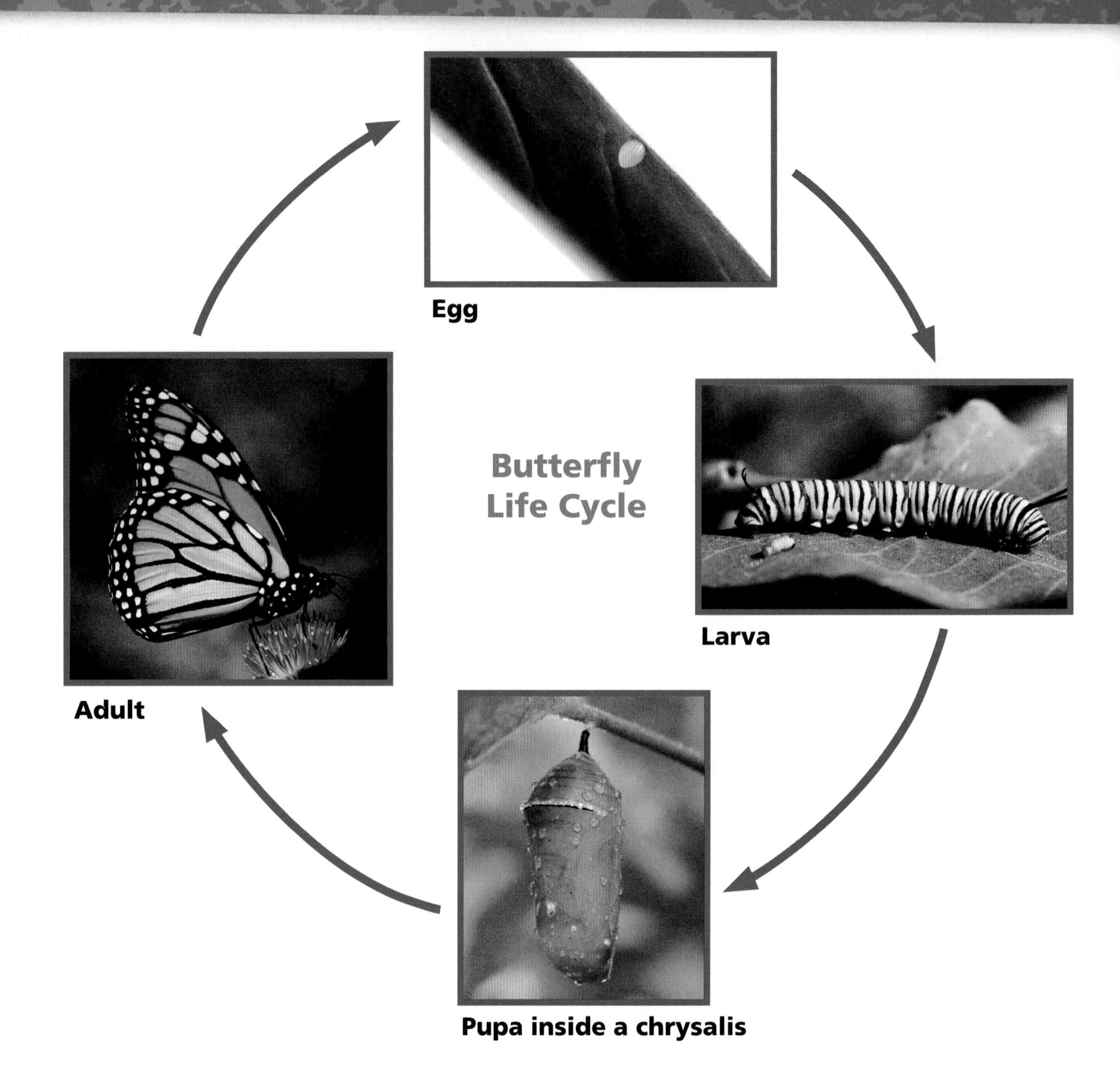

Butterfly Life Cycle

Other organisms have life cycles, too. But animal life cycles can be very different from the life cycle of a bean plant. Some animals are born alive, and some animals hatch from eggs. They all grow up to be adults. The adults mate and produce offspring. The life cycle of the monarch butterfly starts with an **egg**. A tiny larva called a caterpillar hatches out of the egg. The caterpillar eats and grows. When it is about as big as your finger, the caterpillar changes into a **pupa** inside a **chrysalis**. In a couple of weeks, the adult butterfly breaks out of the chrysalis and flies away. After mating, the female lays eggs, completing the life cycle.

Ladybug Life Cycle

Ladybugs, like monarch butterflies, are insects. Ladybugs and butterflies have similar stages in their life cycles. This life cycle is similar to a number of other kinds of insects.

The ladybug life cycle starts when adult ladybugs mate and the female lays eggs. When an egg hatches, a larva comes out. The black larva is the offspring, but it doesn't look like its parents. The larva eats and grows for 3 or 4 weeks before it pupates. Inside the pupa, the larva is changing. When the pupa opens, an adult ladybug comes out. Adult ladybugs are red with black spots. Now the ladybug offspring looks just like its parents. After mating, the female will lay eggs.

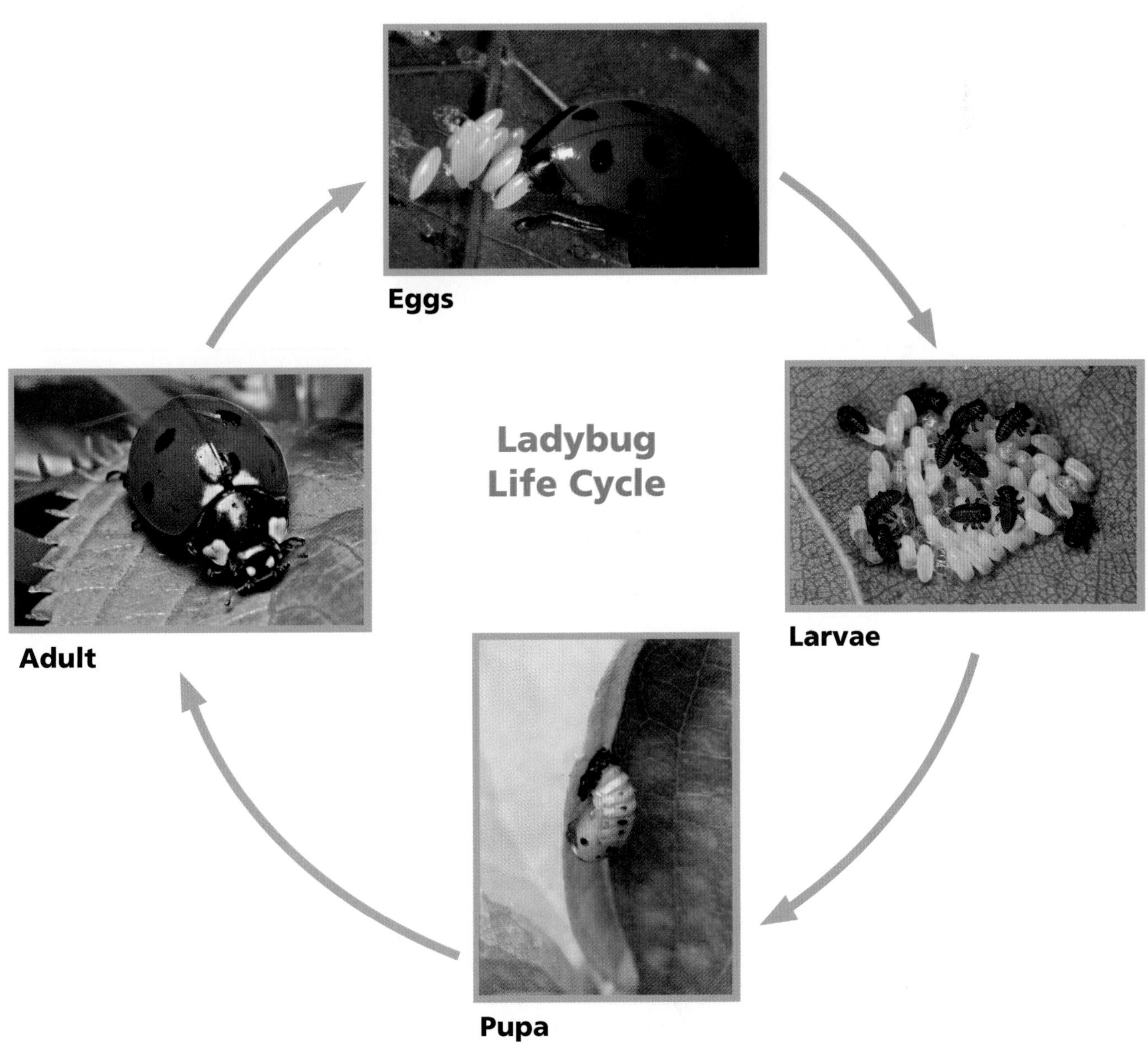

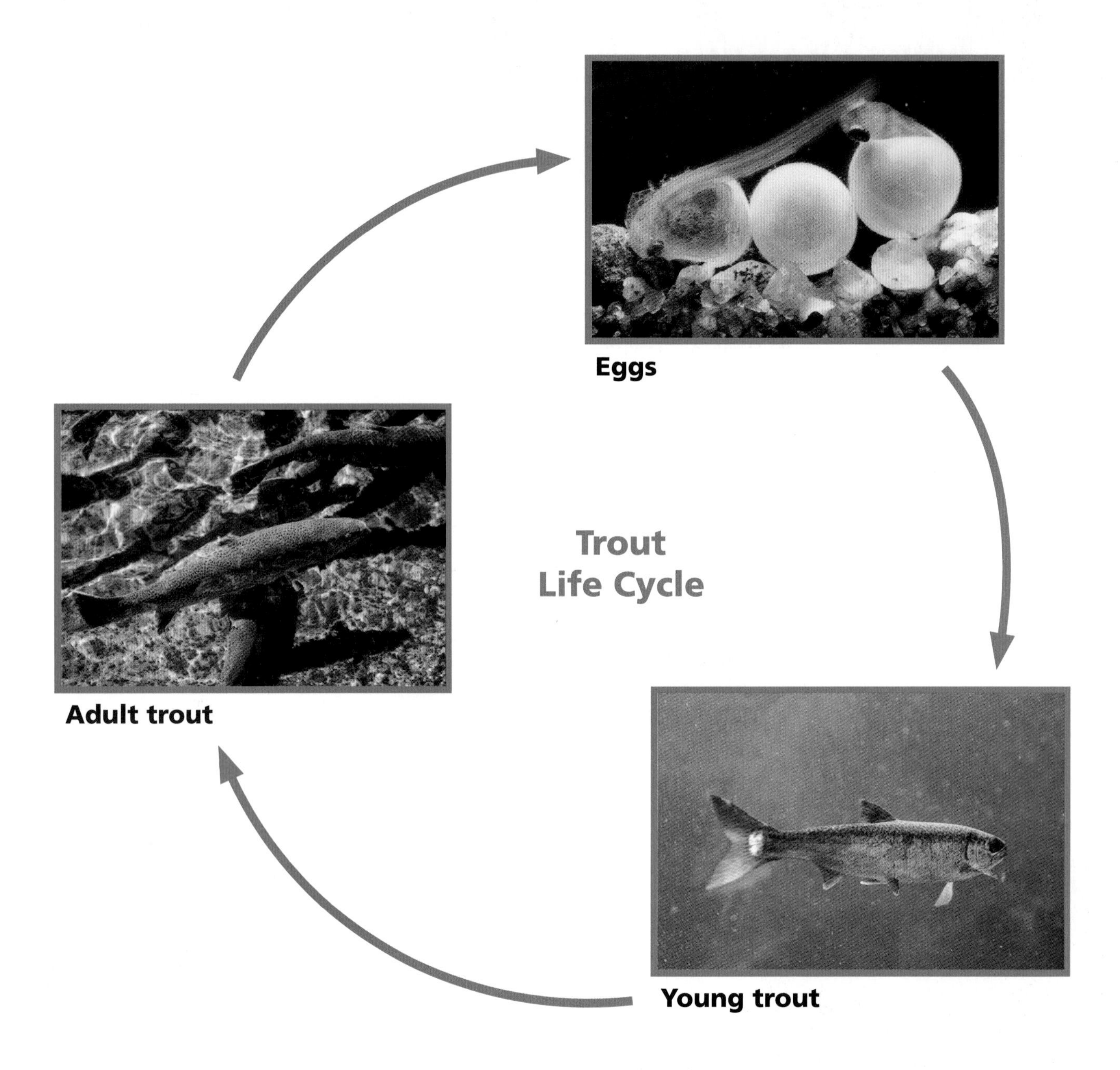

Trout Life Cycle

Trout lay eggs in streams. After 3 to 4 weeks, the eggs hatch, and babies swim out. The babies stay attached to their yolks. The yolks are used as food for another 2 to 3 weeks. Now they are young fish called fry. But they don't look like their parents yet. For the next year, they grow up little by little. In 2 years, they are adults. They look just like their parents. They mate and lay eggs in the stream. Can you describe the trout life cycle?

Frog Life Cycle

Frogs lay eggs in water, too. When an egg hatches, a tadpole swims out. It looks more like a fish with a big head than a frog. A tadpole doesn't look like its parents. The tadpole eats and grows. In a few weeks, the tadpole starts to change. Its long, flat tail gets shorter, and its legs start to grow. In a few more weeks, the tadpole has grown into a frog. Now it looks just like its parents.

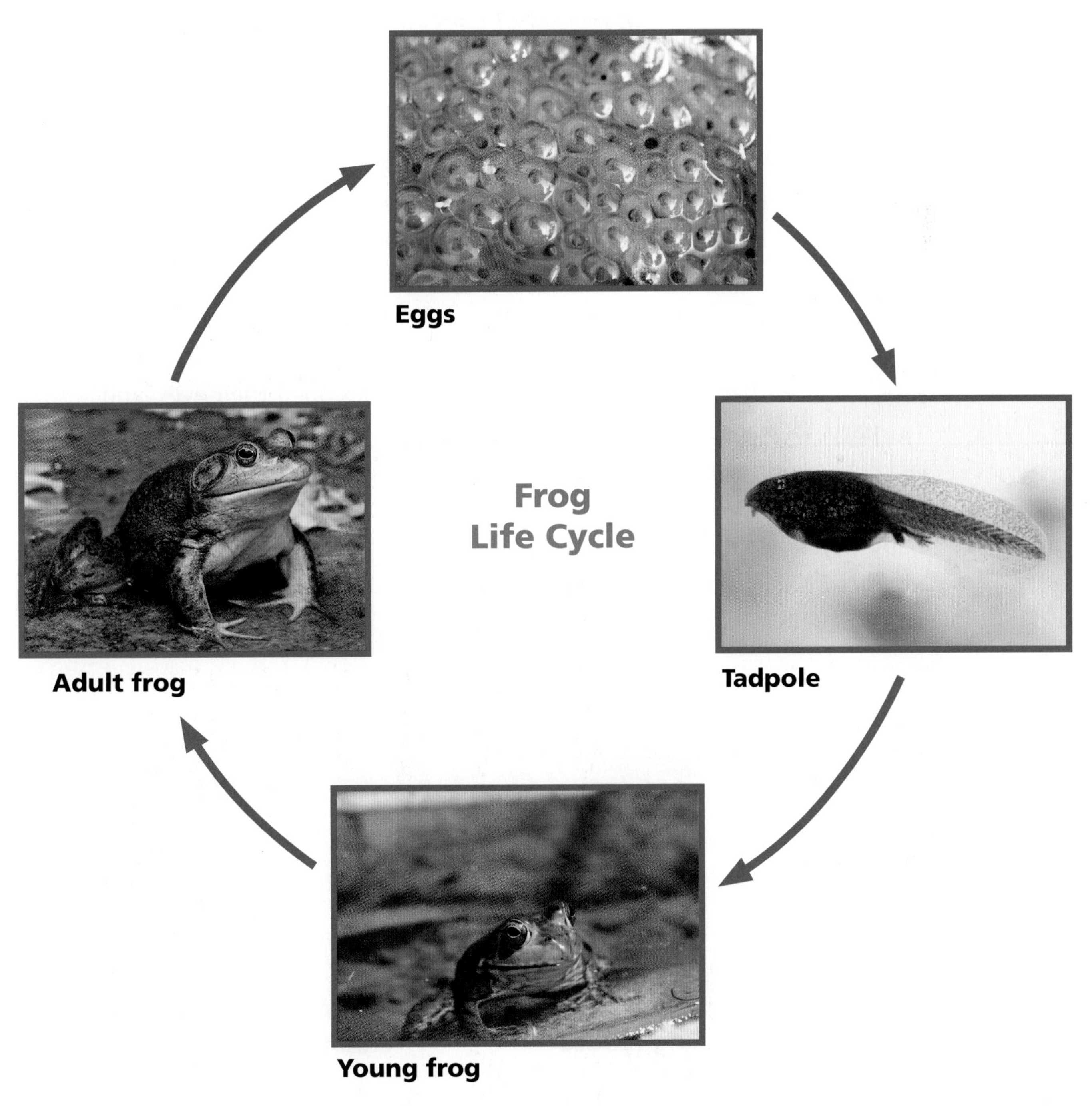

Goose Life Cycle

Eggs **Young geese (goslings)** **Adult goose**

Other Animal Life Cycles

The goose's life cycle starts with an egg. When the egg hatches, a baby gosling comes out. Soon, the offspring grows and matures. In a year, the female goose is ready to mate and lay eggs. The life cycle is complete.

Mammals, such as mice, do not lay eggs. Baby mice grow inside the mother just like humans. The offspring are born alive. Newborn mice are pink, hairless, and blind. You can see that they are mice, but they don't look like their parents yet. In a few days, the babies open their eyes, and fur starts to grow. In a few weeks, the offspring will be adults. They will be ready to continue the life cycle and have babies of their own.

The elephant's life cycle starts with the birth of a baby elephant. The baby elephant eats and grows for years. When a female elephant is 12 or 13 years old, she will mate and have her first baby. With the birth of her baby, the life cycle is complete.

Mouse Life Cycle

Baby mice **Young mouse** **Adult mouse**

Life Is a Repeating Cycle

Plants, insects, fish, frogs, birds, mammals, and all other living things have life cycles. An organism's life cycle is defined by stages. The organism goes through these important stages between the time it is born and the time it produces offspring. The life cycle of the bean takes a few weeks. The life cycles of some insects and the frog take about a year. The life cycle of the elephant takes more than 10 years. All these life cycles are different. Think about the time the life cycle takes for each organism and the stages the organism goes through. Both the time and the stages are different for every different kind of organism.

Thinking about Life Cycles

1. What is a life cycle?
2. Describe the life cycle of one kind of animal.
3. Look at the tomato plant. Describe the life cycle of the tomato.

Crayfish

What do you think it would be like to have your skeleton on the outside of your body? You can study crayfish to find out! Crayfish are members of a group of animals called **crustaceans**. Crustaceans have their skeletons on the outside of their bodies. Crustaceans include lobsters, shrimps, crabs, and crayfish.

Crayfish are aquatic organisms. They live in freshwater environments, such as ponds, lakes, and streams. Crayfish have structures and **behaviors** that let them do many things. Crayfish move around in their environment, get food, protect themselves, and produce offspring. Let's take a look at the interesting structures first. Then we'll find out how those structures help crayfish survive in their environment.

A shrimp **A crab**

What Are All Those Parts on the Crayfish?

The main part of the body is the hard shell called the **carapace**. The head is at the front of the carapace, and the jointed tail is at the back. The legs are attached under the carapace. There are eight small walking legs and two big legs called **pincers**. You can also see two eyes and two long **antennae** at the front of the head. Look for the short antennae in the photographs of the crayfish on the next few pages.

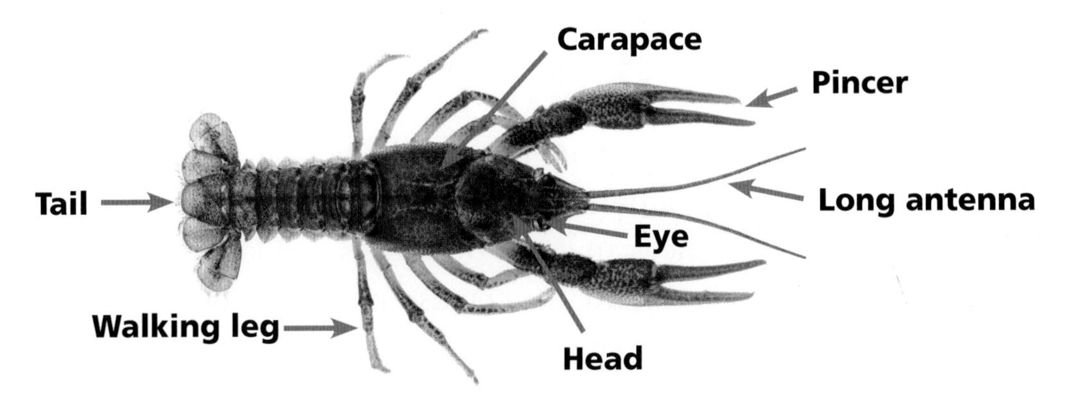

To tell male crayfish from female crayfish, you have to look on the underside. The small, soft legs under the tail, called **swimmerets**, are important. If the first two swimmerets are long with white tips, the crayfish is a male. The male's other swimmerets are short. The female has long, featherlike swimmerets. Also, the female has a white circle between the four back walking legs. This is the egg pore. When the female lays eggs, this is where they come out.

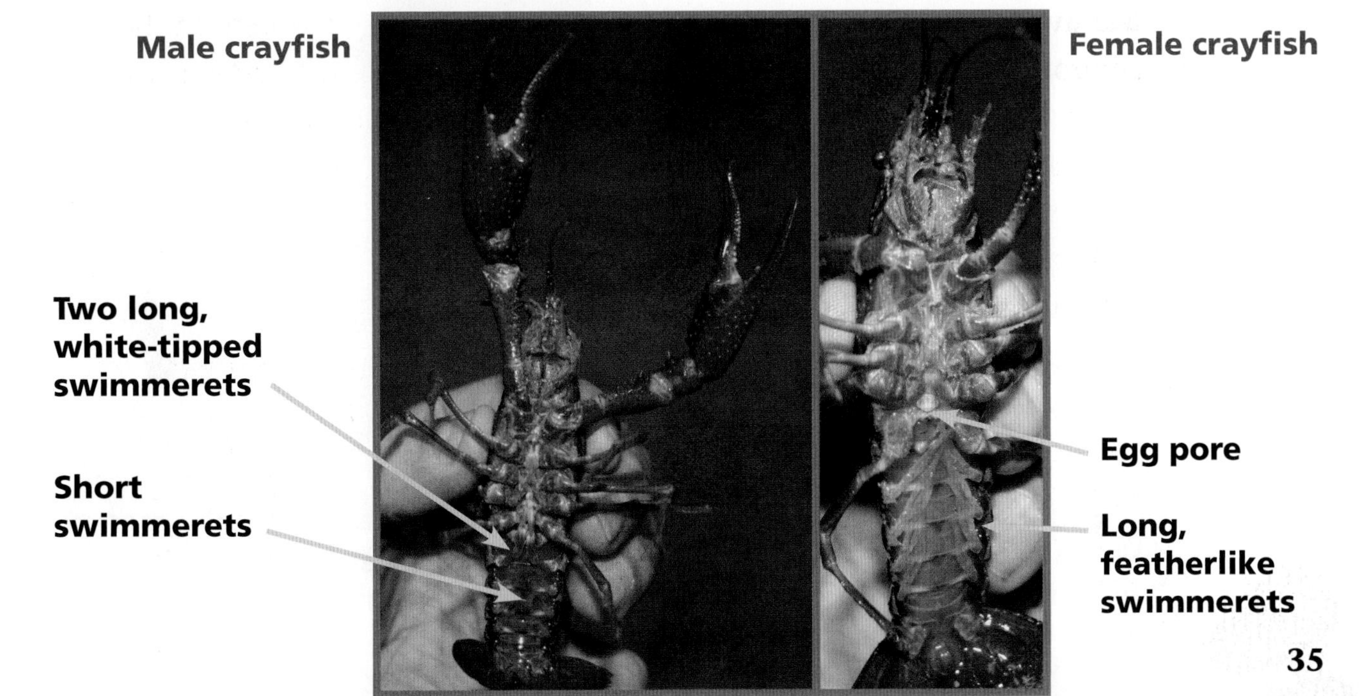

Crayfish have jointed walking legs.

How Do Crayfish Move?

Crayfish spend most of their time on the bottom of a pond or other body of fresh water. To get around, they walk on their eight walking legs. These legs are jointed, so they can climb over rocks and logs easily. Crayfish can walk forward and backward. If a crayfish needs to move quickly, it can shoot through the water very fast. It uses its tail like a paddle. With a quick snap, the tail folds under its body, and the crayfish zooms away backward. This is what crayfish do to escape a predator, such as a large fish or raccoon.

How Do Crayfish Get Food?

Crayfish eat mostly dead plants and animals. They find their food with their antennae, using a sense similar to smelling. Their pincers work well for tearing large food into smaller pieces. And have you noticed that the first four walking legs also have tiny pincers at their tips? These are good tools for picking up small bits of food. The crayfish has interesting mouthparts. They act like scissors and teeth for cutting and breaking food into pieces small enough to swallow.

Crayfish "smell" food with their antennae.

Crayfish can also use their pincers to catch other live organisms for food. The curved points at the tips of the pincers can catch and hold a fish that comes too close.

A crayfish in its hiding place

How Do Crayfish Protect Themselves?

Crayfish have a hard shell that is covered with points, bumps, and bristles. The shell is a little like a suit of armor. This keeps crayfish safe from many predators. Crayfish also use their pincers for defense. When a crayfish is threatened, it raises its pincers as a warning. If the threat continues, the crayfish will use its pincers to attack.

Because crayfish find food with their antennae, they prowl around mostly at night. During the day, crayfish find a place to hide under a rock or log. By being active at night and hidden during the day, crayfish are harder for predators to find.

How Do Crayfish Raise Offspring?

Crayfish start life as an egg. After a male and female crayfish mate, the female lays eggs. However, she doesn't lay them in a nest or under a plant or rock. She carries them under her tail. The long, featherlike swimmerets hold the eggs and fan water around them. She can have 100 eggs or more under her tail.

The eggs start to hatch in 4 to 6 weeks. The babies are only as long as the letter *L* on this page. And they are hard to see because they are transparent. After they hatch, the babies stay under their mother's tail for protection. In a few days, they start to walk around on the gravel. But if they are startled, they scoot back under their mother's tail as quick as a flash.

When the baby crayfish are about 2 weeks old, they leave the protection of their mother's tail. They are ready to start life on their own. When the offspring are 4 to 6 months old, they can mate and produce offspring.

A female crayfish carries her eggs under her tail.

This blue crayfish just molted. It is larger than its old shell.

How Do Crayfish Grow?

Crayfish are completely covered by a hard shell. The shell cannot grow. So how does a crayfish that is less than 1 centimeter (cm) long get to be 10 cm long? The crayfish **molts**.

During molting, the shell splits between the carapace and the tail. Then, with a couple of flips and shakes, the crayfish slides out of its old shell. The crayfish comes out with its new shell already on. But it is soft and flexible. This is when the crayfish grows. Within minutes, it expands. The crayfish is much larger than it was before it molted.

It is important for the freshly molted crayfish to stay hidden. It cannot defend itself or find food when its shell is soft. In 2 days the shell will once again be hard and strong.

A crayfish will molt 6 to 8 times during its life. It may molt 5 times in the first 2 months of life. This is when the crayfish is growing fastest. After its first year, it will molt less often because it is not growing as fast.

Classroom Crayfish

Crayfish are easy to keep in the classroom. All they need is clean, cool water, food to eat, and a place to hide. By observing closely, you can see how they use their antennae to sense their environment. You can see how they use their pincers to defend themselves and to get food. You might be able to see the several mouthparts working as they eat. And you might see them using the small pincers on their walking legs to clean their antennae. You can learn a lot about how their structures and behaviors help crayfish survive and grow in their aquatic environment.

A crayfish in an aquarium

Being Environmentally Responsible

Crayfish are wonderful organisms to study in the classroom, but they can cause problems if they are released into local outdoor environments. The rule is that you never release classroom crayfish or any other organism into natural areas. And if you collect native crayfish from local ponds, you should return them to exactly the same pond, and not move them to another body of water.

Why is this important? There are about 380 different species, or kinds, of crayfish in North America, more than on any other continent. Each kind of crayfish lives in a particular freshwater environment. When an organism is found naturally in an area, it is native to that region. The classroom crayfish may not be native to your region. If an organism isn't naturally found in an area, it is nonnative to that area.

Be responsible when studying crayfish!

A blue crayfish

Sometimes, people introduce nonnative organisms to an area, either intentionally or by accident. Nonnative crayfish used as fishing bait, pets, or science projects should never be released. The introduced crayfish can cause problems by eating the native plants and competing with native animals for food and shelter. Introduced crayfish can eat native animals, including insects, snails, tadpoles, frogs, baby turtles, fish eggs, fish, and snakes. And if there are native crayfish, the nonnative species may **endanger** the native species. Over time, the local crayfish might be entirely replaced by the introduced species.

If an introduced organism thrives in a new area and causes problems, it is called an **invasive** organism. Invasive organisms are changing ecosystems all over the United States. It is important to know how invasive species are introduced and how to prevent their spread.

So remember to do your part to protect your local environment. Never release classroom organisms into local areas.

Thinking about Crayfish

1. What structures help crayfish move around in their environment?
2. What structures help crayfish get food in their environment?
3. What structures and behaviors allow crayfish to defend and protect themselves in their environment?
4. What structures and behaviors allow crayfish to successfully raise offspring in their environment?
5. How might one kind of crayfish become an invasive organism?
6. Find out if there are any invasive plants or animals in your area. What is being done to prevent their spread?

Adaptations

What do porcupines, sea urchins, and cacti have in common? Not the environment in which they live. Porcupines live in the forest, sea urchins live in shallow ocean water, and cacti live in the desert. The answer is that they all have spines. And they have those spines for the same reason. Spines improve the organism's chances for survival.

Any structure or behavior that improves an organism's chances for survival is an **adaptation**. All organisms are able to survive, reproduce, and grow in their environments because they have adaptations.

You already know about several adaptations that crayfish have. Having pincers is an adaptation that helps the crayfish get food. A hard shell is an adaptation that helps the crayfish defend itself against predators. Molting is an adaptation for growing. Carrying eggs under the tail is an adaptation that improves the chances of raising offspring. Those are just a few of the adaptations that crayfish have for living in their environment.

A porcupine

A sea urchin

A barrel cactus

A fish's tail and fins help it swim.

A snake's strong muscles help it move.

A grasshopper's jumping legs help it jump long distances.

Adaptations for Movement

Most animals move in their environment. They need to find food, escape predators, and find mates in order to survive.

Birds fly. Wings and feathers are structures that allow birds to fly. Wings and feathers are adaptations.

Fish swim. Fish have broad tails and fins to move them through the water. Fish have a streamlined shape. Broad tails, fins, and a streamlined shape are adaptations that allow fish to move easily through their environment.

Snakes slither. Snakes have strong **muscles** that make waves along their bellies. They have scales that give the snake traction. The waves push the snake forward. Strong muscles and scales are adaptations that allow snakes to move through their environment.

Grasshoppers walk, jump, and fly. They have walking legs for moving slowly through the grass. Grasshoppers have strong legs for jumping long distances and wings for flying. Walking legs, jumping legs, and wings are adaptations that allow grasshoppers to move through their environment in three different ways.

Any structure or behavior of an animal that allows it to move in its environment is an adaptation for movement. What adaptations do you have for moving in your environment?

These animals have different structures to help them get food.

Adaptations for Getting Food

Animals can't make their own food. They have to find and eat food to survive. Every animal has structures and behaviors for getting the food it needs to survive in its environment.

Frogs eat insects. Frogs have long tongues with a sticky pad on the end. The frog shoots out its long tongue at an insect. The insect sticks to the pad. The long tongue and sticky pad are adaptations that allow frogs to catch insects to eat.

Barnacles don't move to get their food. They wait for food to drift by. Barnacles have specialized rakes they wave in the water. Small organisms get caught in the rakes. Specialized rakes are adaptations that allow barnacles to get the food they need to survive in their environment.

Woodpeckers eat insects in trees. They have strong, sharp beaks and strong neck muscles. Woodpeckers chip away bark and dead wood to find the insects they eat. Sharp, strong beaks and strong neck muscles are adaptations that allow woodpeckers to get food in their environment.

Butterflies eat nectar from flowers. To reach into deep, narrow flowers, a butterfly has a long, strawlike mouth called a **proboscis**. The proboscis is an adaptation that allows the butterfly to get food.

Any structure or behavior of an animal that allows it to get food in its environment is an adaptation for feeding. What adaptations do you have for getting food?

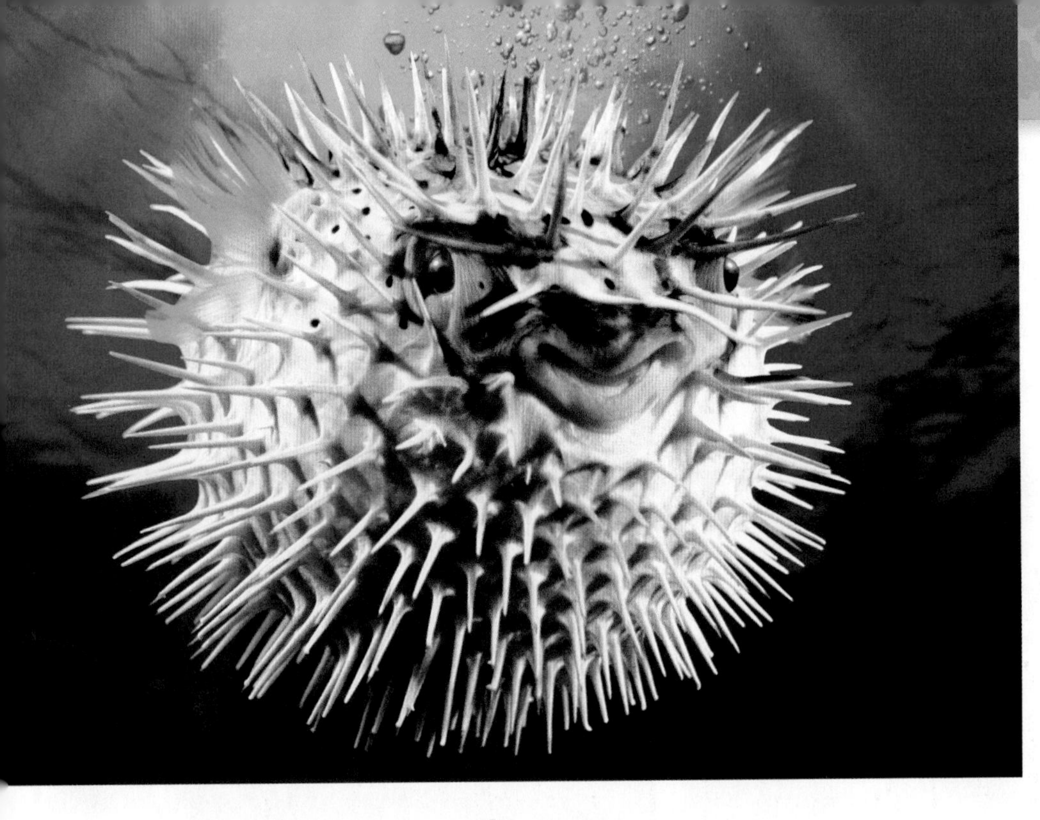

Spines help protect this spiny puffer fish.

Adaptations for Protection

Both plants and animals need to protect themselves from predators and weather. Every successful plant and animal has adaptations for defending itself.

Spiny puffer fish are small and swim slowly. They would make an easy meal for a larger fish. But this kind of puffer fish is covered with structures called spines. When it is threatened, the fish puffs up. Spines and the ability to puff are adaptations that protect this spiny puffer fish. These structures and puffing behavior allow the fish to survive in its environment.

Butterflies don't have spines. They can't fly fast. But some of them have colors and patterns that help them blend in with their environment. Blending in is called **camouflage**. Camouflage is an adaptation that protects this butterfly from predators in its environment.

This butterfly is camouflaged to look like a dead leaf.

This tortoise is protected by its hard shell.

Poisonous sap protects this milkweed plant from being eaten.

This sea otter has waterproof fur to keep it dry and warm.

Tortoises stay safe by wearing armor. Their hard shells are difficult for a predator to break into. A hard shell is a very effective form of protection. A hard shell is an adaptation to keep the tortoise safe from predators.

Milkweed plants have poisonous sap. Most animals that try to eat milkweed plants get sick. Poisonous sap is an adaptation that protects the milkweed plant from being eaten by hungry animals.

Sea otters live in the cold coastal waters of North America. The sea otter's dense fur traps a layer of air. Water cannot get to the sea otter's skin. It stays warm in the icy water. Thick, waterproof fur is an adaptation that protects the sea otter from the cold in its environment.

Adaptations for Reproduction

Dandelions produce hundreds or thousands of seeds.

Every kind of plant and animal must reproduce. Every organism has adaptations that allow it to produce offspring. Some organisms spend a lot of time raising and caring for their offspring. Other organisms spend no time raising offspring. The method of reproduction that works for one kind of organism will not work for another kind of organism. Every organism has its own adaptations for reproduction.

Dandelions are successful plants. Each plant produces hundreds or thousands of seeds. Each seed has a puff of down to carry it on the wind to a new location. Once the seeds blow away, they are on their own. The dandelion's adaptation for reproduction is to produce many seeds.

The grebe is a waterbird. The female lays eggs in a nest in a marsh. When the chicks hatch, they follow their mother as they find food. When they get tired, they climb on her back and snuggle down under her feathers. The behavior of protecting the chicks is an adaptation that helps the baby grebes survive.

This grebe protects her chicks by hiding them under her feathers.

Bees are social insects. They live in colonies of thousands of workers. Workers build six-sided wax cells. The queen bee lays one egg in each cell. When the eggs hatch, the worker bees feed and care for the growing larvae. When the larvae are fully grown, the workers cover the cells with wax. In a few days, the adult bees come out, ready to go to work. Feeding and caring for the young is an adaptation that improves the bee colony's chances of survival.

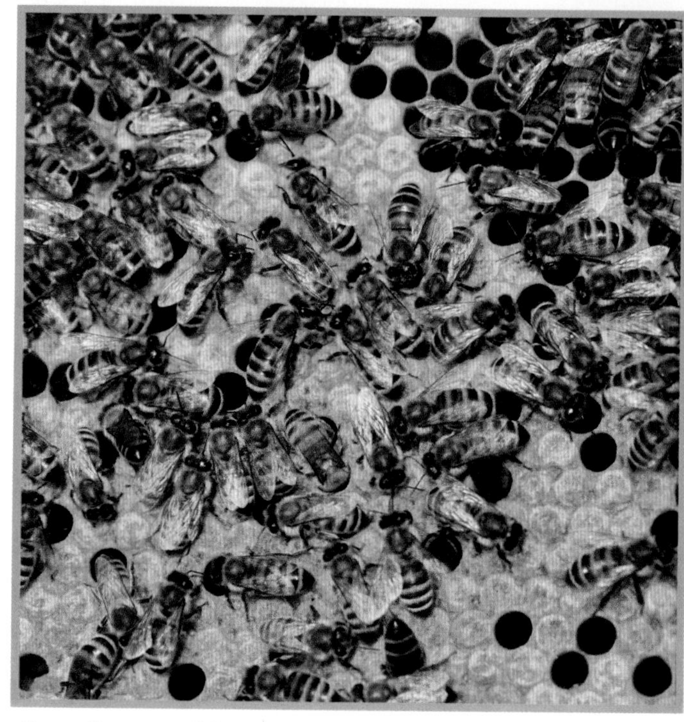
A colony of bees

Human babies are helpless when they are born. They grow and learn slowly. Human parents must spend years raising their offspring before the offspring are ready to go out on their own. Providing years of support and care is an adaptation that improves the chances that human offspring will survive.

Humans raise their offspring for years.

What adaptation do you see here?

Organisms are adapted to live in a certain environment. An organism's adaptations don't help if an organism is not in its environment. For example, the barnacle's rake is an adaptation for getting tiny food particles in the ocean environment. The rake helps the barnacle survive. But what if the barnacle is moved to an environment that has only large food particles? The rakes won't work. The barnacle will die because it is not adapted for eating large food particles.

Adaptations make it possible for many different kinds of organisms to live in the same environment. Each different organism has adaptations that allow it to use different resources in the environment.

Thinking about Adaptations

1. What is an adaptation?
2. What are some adaptations crayfish have for survival in their environment?
3. What are some adaptations organisms have for movement in their environments?
4. What are some adaptations organisms have for protecting themselves in their environments?
5. What are some adaptations organisms have for getting food in their environments?
6. What are some adaptations organisms have for successfully producing offspring?

Life on Earth

The diversity of life on Earth is amazing. Organisms live in every environment you can imagine. They live in lakes, on mountains, in swamps, and everywhere else. Some environments are hot, some are very cold, some are wet, and some are dry. The conditions in an environment determine what organisms can and can't live there. Plants and animals that live in a pond do not have adaptations to live in a forest. Plants and animals that live in a forest can't live in the ocean. Plants and animals have adaptations that help them grow, survive, and reproduce in their environments.

Wetlands

Wetlands are places with ponds, streams, swamps, and muddy fields. Cattails and other plants grow in the water. Cottonwood and willow trees line the ponds and streams.

In many wetlands, cattails are the most successful plants. The strong roots hold tightly to the bottoms of ponds. The brown hot dog-shaped structures are seed heads. Each seed head has thousands of seeds.

A catfish

A bullfrog

Many animals live in the water. Crayfish live on the bottoms of ponds. Their hard shells and strong pincers help them defend themselves from predators. Their wide tails let them swim away quickly. The females carry their eggs under their tails in long swimmerets. There, the eggs are safe and more likely to survive and grow up to become adult crayfish.

Other animals share ponds with crayfish. Catfish have adaptations that allow them to live in dark and murky water. They look for insects, crayfish, and bits of plants to eat. Catfish use their whiskers to sense their environment and find food. Their fins let them move easily through water. They have two sharp spines on their sides for defense. Catfish live their whole lives in the water.

Bullfrogs live near the edges of some ponds. They have long tongues with sticky pads on the end for capturing large insects like dragonflies. They have strong legs for jumping and webbed feet for swimming. Bullfrogs jump into the water to escape land predators.

Raccoons can live in the willows and cottonwood trees near wetlands. They come down to the ponds to look for food. They might catch a frog or crayfish, or find a bird's nest with eggs. Their handlike front feet and sharp teeth allow them to eat many different things.

Raccoons

Each fall, wetlands across the southern part of the United States have visitors that stay from October to April. The visitors are thousands of ducks, geese, and cranes. The birds fly south from their breeding grounds in Canada and Alaska. They are known as waterfowl.

The wetlands provide food and safety for waterfowl during the winter. The birds swim calmly on the ponds and fly out to fields and marshes to eat seeds. The weather is mild, and there are few predators. As long as the wetlands have water to fill the ponds and flood the fields, waterfowl will continue to thrive.

Geese in a wetland

Deserts

Deserts are very different from wetlands. Deserts are dry, often hot, and rocky. Desert plants have adaptations for getting and holding water. Cacti have thick stems to store water. Desert trees have small, waxy leaves to save water.

A desert in Arizona

A cactus wren

Desert animals need water, too. Cactus wrens can fly to water for a drink. Desert coyotes can travel long distances on long legs to get water. Kangaroo rats and desert tortoises get water from the seeds, leaves, and flowers they eat. They rarely need to drink water.

A desert coyote

A desert tortoise

A pine and fir forest in the mountains

Forests

Pine, fir, spruce, and hemlock trees thrive in northern forest environments. Mountains are cold in the winter and warm in the summer. The forest trees must be able to survive months of cold winter with deep snow on the ground.

The snowshoe hare has adaptations that allow it to live in the forest all year. In winter, it grows white fur. The white fur blends in with the white snow, making the hare hard to see. Camouflage improves the hare's chances of survival. The sharp-eyed owl can spot the hare only when the hare moves. The owl's sharp talons and strong beak allow it to catch the hare for a meal.

A snowshoe hare

A great horned owl

The black bear has strong claws for digging up roots and tearing apart dead trees to find food. When winter comes, the bear finds a den for shelter. It will **hibernate**, living off its layer of fat. In the spring, when the snow is gone, the bear comes out to feed on the fresh berries and plants.

A black bear

Deer are the fastest-running animals in the forest. They are always alert for danger as they nibble grass and twigs for food. When they sense danger, they can run away in a flash. Deer leave the high forest in the winter and move to lower areas where they can find food.

A mule deer

A grassland

Grasslands

The prairie and the Great Plains of North America are grasslands. As you might expect, grasses are the main plants growing there. Few trees and bushes grow in the grasslands. Range fires often sweep across the grasslands, burning the dry grass and small trees. Grass plants are not killed by fire. They can grow new blades from their underground roots in the spring. Many other plants are not adapted to survive fire. Fire helps the grasslands stay the way they are.

Many animals live in the grasslands. Prairie dogs have strong legs and claws for digging tunnels underground. They come out to eat the grass. Grasshoppers live right in the grass. They jump and fly from place to place to get food. American bison wander across the grasslands, eating the grass as they go. Horned larks are seasonal visitors. They **migrate** to the grasslands in the summer to feed on grass seeds and raise their young. In the winter, horned larks migrate south where it is warmer.

A horned lark

An American bison

A prairie dog

Fire is a challenge for animals living in the grassland environment. Large animals like bison can move to a new location to escape the fire. Prairie dogs can retreat into their tunnels. Horned larks can fly away. But if the horned larks have eggs in a nest or babies that can't yet fly, they will die. The grasshoppers are not strong fliers. They can fly a short distance to safety, but if the fire is large, the grasshoppers will die.

In the spring after a fire, new grass sprouts come up from the roots. The ashes provide nutrients for the new grass. The animals that ran from the fire will return. Grasshoppers will fly in and reproduce. The grassland is soon full of life once again.

A grasshopper

A tundra in summer

Tundra

The tundra is cold, frozen land most of the year. Northern Alaska is a tundra. During the winter, the ground is frozen. Days are short. Plants stop growing, and most animals seek shelter from snow and wind.

Only animals with thick fur or feathers can survive the tundra winters. Arctic foxes scavenge for scraps of food. Ptarmigans scratch through snow to find seeds and small plants. Foxes and ptarmigans grow white fur and feathers in winter. They blend in with the white environment.

An arctic fox

A ptarmigan

In the summer, days are long, and the weather is warm. The soil defrosts. The tundra comes to life. Millions of mosquitoes swarm over the pools and marshes where they reproduce. Millions of birds come to the summer tundra to raise their young. Summer is also when the snowy owl raises its offspring. The owl catches mice and voles to feed its young. It also catches baby birds and fish from time to time.

The tundra goes through big changes in weather over the year. Animals that don't have structures to protect themselves from the cold have to leave after the summer to survive.

A snowy owl

A mosquito swarm

Ocean and Ocean Basins

More than half of Earth's surface is ocean. Water in the ocean is cold in the far north and far south, and warm in the middle parts of the planet. The warm ocean basins are called tropical oceans.

Life on land is based on plants. But there are few plants in the tropical oceans. The tropical oceans are known for the diversity of their animals. The most important animals in tropical oceans are the corals. They surround themselves with hard shells for protection. In this way, they build huge structures called reefs. Thousands of animals make their homes in the coral-reef environment.

A stingray

A butterfly fish

The corals build huge reefs that provide many places for other animals to grow, hide, and find food. The reef benefits animals that need a surface to stick to, like snails and clams. The reef benefits animals that need places to hide, like shrimp and lobsters. The reef provides a resting place for ocean travelers like the green sea turtle.

Some ocean organisms live part of the year in cold ocean waters and other times of the year in warm tropical ocean waters. Humpback whales are found in the ocean and seas around the world. They migrate up to 25,000 kilometers (km) each year. They feed on krill and small fish in the summer in the cold polar environment. In the winter, the whales migrate to tropical waters to breed and give birth. They don't eat during that time but live off the fat in their bodies built up during the winter. Adult humpback whales are very large mammals. They are 15 to 17 meters (m) long. Young whales, called calves, are about 6 m long. Humpback whales are adapted to live in both polar and tropical waters at different times of the year.

A humpback whale

Organism Diversity

A crayfish has adaptations to survive in fresh water.

The crayfish is an example of a successful organism. It has adaptations that allow it to live in lakes, ponds, and streams. The crayfish has gills for getting oxygen from water. It has pincers for getting food. It has a broad tail for fast movement through water to escape danger. Crayfish carry their eggs under their tails to make sure the eggs are safe while they develop. Crayfish have antennae to find food in the dark. All of these adaptations make it possible for crayfish to survive, grow, and reproduce in aquatic environments.

The crayfish's adaptations would not help it survive in a grassland. Its gills would not get oxygen from the air. It doesn't have strong claws to dig a tunnel for shelter. Crayfish can't run fast or fly to escape range fires. Crayfish are not adapted to live in grasslands. Prairie dogs, horned larks, and bison are adapted to live in grasslands. Organisms live in environments for which they have adaptations.

Thinking about Life on Earth

1. What kinds of organisms live in wetlands?

2. What kinds of organisms live in grasslands?

3. What kinds of organisms live in the ocean?

4. What kinds of organisms live in forests?

5. What kinds of organisms live in the tundra?

6. What kinds of organisms live in deserts?

Inside a Snail's Shell

A snail has many of the same body parts as humans, but they are in different places. A snail's teeth are on its tongue. A snail's toothy tongue is called a radula. Did you know a snail's mouth is on its foot? And its breathing hole is next to where it excretes waste!

A snail has several tentacles on its head. Its eyes are at the ends of the two longer tentacles. A snail doesn't have very good eyesight, but it can sense light from dark.

A snail also uses its tentacles to feel its way around. Scent detectors on the tentacles help the snail find food. The tentacles also tell the snail when other animals are close by.

A snail's foot is covered with tiny stiff hairs called cilia. These help the foot grip the ground. A gland in the snail's foot produces a thick trail of slime to help the snail slide along.

A snail's hard shell protects it from predators. It also gives the snail a safe place to stay when it is hot and dry outside. The shell becomes thicker and harder as the snail grows. It coils around the snail's body as it grows.

Snail shells have a variety of designs.

Snail Facts

A cone snail is armed with a poisonous stinger.

- Snails have shells. Their close relatives, slugs, do not. The shell is the main difference between snails and slugs.
- Snails are part of the class called **gastropods**. That word is from two Greek words. *Gastro* means "belly." *Pod* means "foot."
- A snail's tongue can have as many as 150,000 teeth.
- Snails produce slime to help them move.
- A snail makes its shell bigger by adding new shell material around the opening of the shell.
- There are about 40,000 different kinds of snails.
- Snails live on land, in fresh water, and in the ocean.
- The largest land snail is found in Africa. Its shell can be 25 centimeters (cm) long!
- Most land snails are both male and female. That means any snail can produce eggs.
- Land snails lay from 30 to 50 tiny eggs in a hole in the ground. The snail doesn't stay to protect the eggs, so many are eaten by insects.
- A snail's shell grows for the first 2 years of its life. By then, the shell can have four or five coils.
- Snails are right-handed or left-handed. If a snail's shell coils to the right, it is right-handed. A left-handed snail has a shell that coils to the left.
- Most snails live about 3 or 4 years.

A jewel top snail

A Change in the Environment

All plants and animals change the environment in which they live. Trees create shade. Flowers produce odors. Squirrels dig burrows in the soil. Woodpeckers drill holes in trees. And every animal eats something. Most changes to the environment are small. Most plants and animals living there continue to go about their business.

Some animals, however, change their environment a lot. As a result, many animals living in the environment must move or die. Plants cannot move, so if the changes are bad for them, they die. One animal that makes big changes to its environment is the beaver.

Beavers live around water. They build a mud-and-stick lodge right in the water. The entrance to the lodge is underwater. The part of the lodge where they live and raise their young is above water. Beavers are safe and comfortable in their lodges.

If a lake or pond is nearby, the beaver family makes its lodge there. If there is only a stream, the beavers build a dam to make a pond.

A mud-and-stick lodge

A pond created by a beaver dam built across a stream

First they use their large, sharp front teeth to cut down trees by the stream. They cut off the branches and drag them into the stream. They put mud and rocks on the branches to hold them in place. Then they add more branches and mud.

The beavers are making a dam. They keep adding to the dam until it reaches all the way across the stream. Water is trapped to form a pond. The beavers then build their lodge.

So what about the plants and other animals living in the stream environment? Some of them benefit from the beaver's work. Others can't live there anymore. The beavers cut down the trees for food and building material. This means less shade. When water floods the land around the stream, the grasses, bushes, and trees living there die. Insects, snakes, squirrels, and all the other land animals have to move out. When beavers build a pond, the animals that live on the land by the stream are forced to move, and the plants growing along the shore die.

It's different for the aquatic plants and animals. Fish and frogs have a lot more room to live. Aquatic plants, like cattails and water lilies, thrive. Some aquatic insects, like dragonflies and mosquitoes, benefit from the changed environment. Aquatic plants and animals grow and reproduce to take advantage of the larger environment created by the beavers.

A riparian environment

The Riparian Brush Rabbit

Beavers are not the only animals that change streamside environments. Humans do, too. People build dams on rivers and streams. People also change the environment along rivers and streams when they build towns, houses, and farms.

Changes along rivers and streams are detrimental to one small rabbit in California called the **riparian** brush rabbit. The word *riparian* means "along a river or stream." Bushes, vines, and branches grow thick along rivers and streams. The brush rabbit thrives in this environment.

When people build homes and plow fields for farms, they clear away the brush. When this happens, the brush rabbit has no place to hide. Without protection, it is easily caught by coyotes, raccoons, and hawks.

A riparian brush rabbit in its environment

In 2000, the riparian brush rabbit was listed as endangered. That means it is in danger of dying out. There was only one **population** living in a state park. When the park flooded in 1997, the rabbits were nearly wiped out. Wildlife **biologists** figured out that something must be done to provide a suitable environment for the brush rabbits. If not, all the rabbits could die.

Something is being done. The US Fish and Wildlife Service, the California State University Stanislaus, and a private ranch have joined forces. These three groups are working together to provide a good environment for brush rabbits. A small group of adult rabbits was released at the ranch by the San Joaquin River. The brush is thick. The environment is good for riparian brush rabbits. Everyone is hoping that the released rabbits will have offspring, increase in number, and start a new population.

Thinking about Changes to Environments

1. How do beavers change their environment?
2. What organisms benefit when beavers dam a stream? What organisms suffer?
3. What caused the riparian brush rabbit to become endangered?
4. Is there an organism in your region that is endangered? Find out about the organism and what is being done to protect it.

Food Chains

Every animal depends on other animals or plants to survive. Eating is the way animals get the food they need to survive. What is it about food that makes life possible? Food is a source of matter and energy. The matter in food provides the raw materials an organism needs to grow and reproduce. Energy is like fuel. It makes things happen.

The transfer of energy and matter from organism to organism in a feeding relationship is called a **food chain**. Let's look at the links in a food chain.

Some organisms don't eat anything. They don't have to because they make their own food. On land, plants such as grasses, trees, and bushes make their own food. In freshwater and ocean systems, plants and algae make their own food. Plants and algae use the energy from the Sun to make their own food. Plants and algae are the primary source of matter in a food chain.

Plants make their own food using energy from sunlight.

What Animals Eat

Some animals eat plants and plant parts. Deer eat grass, leaves, and twigs. Gophers eat roots. Squirrels eat nuts and berries. Grasshoppers eat grass. Pond snails and fish eat algae. Animals that eat only plants or algae are called **herbivores**.

Some animals don't eat plants. Snakes don't eat nuts and berries. Hawks don't eat grass. Spiders don't eat leaves. So how do they get their matter and energy? They eat other animals. Snakes and hawks eat gophers and squirrels. Spiders eat insects. Frogs eat insects. Sea otters eat abalone and sea urchins. Animals that eat other animals are called **carnivores**. Carnivores that catch live animals are called predators. The animals they eat are called **prey**.

Some animals, like bears, raccoons, robins, and crayfish, eat both plants and animals. They are called **omnivores**.

Some animals eat dead organisms. Some, like vultures, eat only dead animals. Others, like isopods and termites, eat dead leaves and wood. Coyotes, rats, and ants will eat just about anything that is dead. Crayfish will also eat dead plants and animals. Animals that feed on dead organisms are called scavengers.

A deer is an herbivore because it eats only plants.

A bear is an omnivore because it eats both plants and animals.

Food Chains

When a frog eats a grasshopper, the matter and energy in the grasshopper go to the frog. This feeding relationship can be shown with an arrow. The arrow always points in the direction that the matter and energy flow.

grasshopper → frog

If a hawk eats the frog, the matter and energy in the frog goes to the hawk. Matter and energy pass from one organism to the next when it is eaten. This is the food chain.

grasshopper → frog → hawk

The first link in this chain is the grass. The second link is the grasshopper. The third link is the frog. And the last link in the food chain is the hawk. The arrows show the direction of energy flow. They point from the organism that is eaten to the organism that eats it.

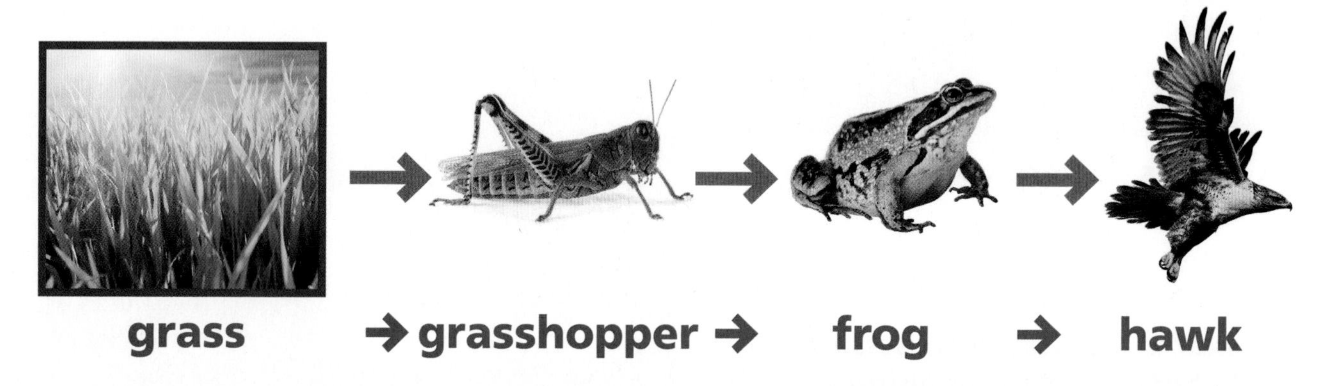

grass → grasshopper → frog → hawk

A food chain is a way to describe a feeding relationship between organisms. Matter and energy transfer from one organism to the next organism in a food chain. Plants and algae make their own food using the energy from the Sun. The Sun is needed for almost every food chain on land or in water.

Plants are the primary sources of matter in many food chains.

Connecting the Links in Food Chains

A population is a group of organisms of one kind that lives in an area. In a field of grass, there might be a population of grasshoppers. In and around a pond, there might be a population of frogs. Around the pond, there might be a population of hawks.

1. What might happen to the food chain if the population of grasshoppers gets larger? What if the population of grasshoppers gets smaller?
2. What might happen to the food chain when a drought or fire destroys the grass?
3. What might happen to the food chain when the hawks fly away to hunt in a different area?

The Human Skeleton

One of the most marvelous systems in the world is the human body. The many parts of your body work together, allowing you to walk, run, jump, and play. Even while you sleep, your body keeps working.

Your skeleton is an important part of you. It is the framework of the body and gives the body its shape. As you grow, your skeleton grows and changes with you. Your bones grow, and some even fuse together. By the time you are 1 year old, your skeleton has about 206 different bones.

Super Protectors

Bones do more than just support the body. They also protect the soft organs inside. Here are some bones that act as armor for your organs.

Skull The skull keeps the brain and sensory organs safe from harm. The skull is made up of 26 different bones. The lower jaw and the tiny ear bones are the only bones in the skull that move. Did you know that your teeth are not bones? They are hard, bonelike structures. Teeth have a hard outer layer and a soft, pulpy inner layer. The outer layer of a tooth is covered by a super strong coating of enamel, which is the hardest substance in the human body.

Ribs In your chest are 12 pairs of ribs. These ribs protect the heart, lungs, spleen, stomach, and liver. As you breathe in, your ribs move up and out. This helps your lungs take in more air.

Pelvis The pelvis is made up of three bones. One is a bone at the base of the spine called the sacrum. The two other bones are the hip bones that make up the pelvic girdle. The pelvic girdle cradles and shields the intestines and the bladder.

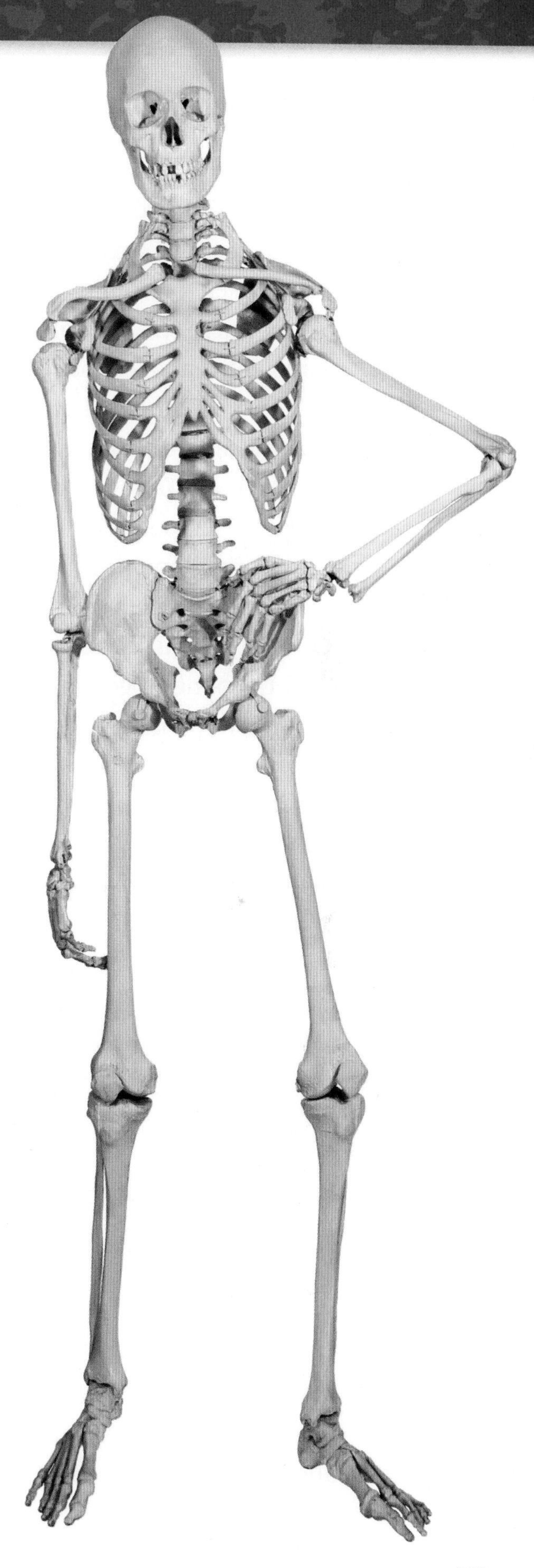

A Flexible Framework

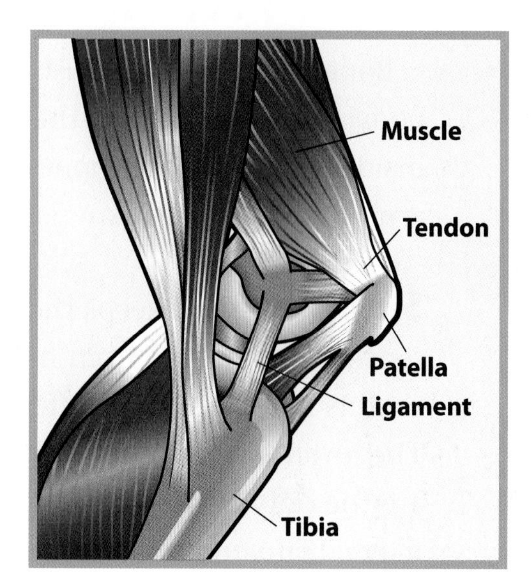

Each bone in the human body is hard and unbending. Yet the skeleton itself is flexible. **Joints** make it possible for the skeleton to move. Joints are the places where two or more bones meet. Some joints allow a lot of movement. Others move only a little.

Bones are held in place by connective tissues. **Cartilage** is a kind of connective tissue. It is found at the ends of the bones. Cartilage protects the bones and helps joints move smoothly. **Ligaments** are another type of connective tissue. They hold the bones together at the joints.

Bones don't move by themselves. Muscles move bones. Together bones and muscles allow movement to happen at joints. How? **Tendons**, another type of connective tissue, connect muscles to bones. When a muscle **contracts**, or shortens, the tendons pull the bones, causing movement.

Here are some super flexible parts of your body.

Spine The spine is the backbone of the body. The 26 bones in the spine are called vertebrae. They have cartilage between them, allowing the spine to bend and twist.

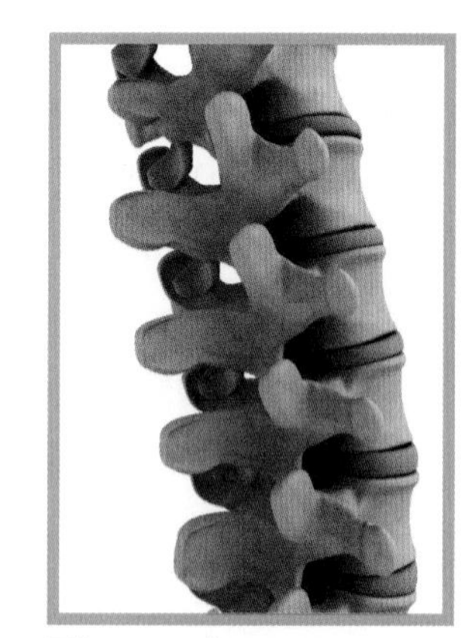

The spine

Shoulder The scapula, or shoulder blade, and the arm are connected by many muscles and ligaments. This flexible shoulder joint allows you to swing your arm in a full circle. Try it!

Hip The femur, or thighbone, is the longest bone in your body. One end of it fits perfectly into your pelvis at your hip. The hip joint allows you to kick to the front and to the side.

What's inside Your Bones?

The bones in your body are living tissue. They are made mostly of calcium and protein. To stay strong, bones need oxygen, vitamins, and minerals. A good diet and plenty of exercise can help keep bones healthy and strong.

A bone is made up of several different parts.

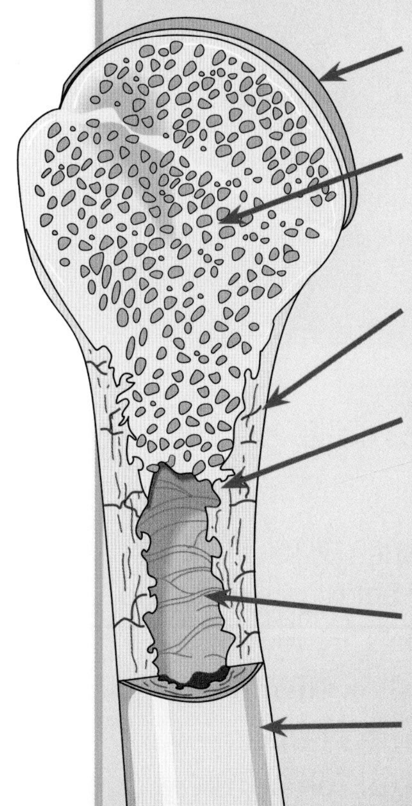

Hard outer layer The hard outer layer gives the bone its strength. It is made of dense, compact bone.

Spongy inner layer The spongy inner layer makes the bone light. This same spongy material makes up the ends of the bone.

Blood vessels Blood vessels bring oxygen and other nutrients to the bone.

Marrow cavity The marrow cavity is the space in the center of the bone. It contains bone marrow. The marrow cavity is found only in long bones such as the leg and arm bones.

Bone marrow Bone marrow is a tissue that makes blood cells. It is found in the marrow cavity of some bones.

Periosteum The periosteum is a thin, tough layer that covers the bone surface.

Name That Bone

Sure, you can touch your kneecap, but can you point to your patella? Check out the scientific terms for some of the bones in your body.

Maxilla upper jaw
Mandible lower jaw
Scapula shoulder blade
Clavicle collarbone
Sternum breastbone
Coccyx tailbone
Humerus, **radius**, **ulna** arm bones

Carpals wrist bones
Phalanges finger bones
Femur, **tibia**, **fibula** leg bones
Patella kneecap
Tarsals foot and ankle bones
Phalanges toe bones

Barn Owls

Barn owls are large birds with white, heart-shaped faces. They are found all over the world. Barn owls live near fields, pastures, and any places where voles and other small animals are found.

Barn owls don't tear or chew their prey. They swallow their prey whole. But they cannot digest the bones, fur, claws, and teeth of their prey. About 20 hours after feeding, barn owls regurgitate, or spit up, these bits. They come up as oval balls of fur and bones called owl pellets. The pellets are from 3.75 to 7.5 centimeters (cm) long.

Barn owls don't tear or chew their prey. For this reason, the pieces of a complete skeleton of a small rodent can almost always be found in a pellet. Sometimes a pellet might contain the remains of several small animals. You can use a toothpick to pull a pellet apart. Then you can put together at least one skeleton of a barn owl's meal from the pellet.

Barn owls aren't the only birds that spit up pellets. All owls regurgitate indigestible parts of their food as pellets. Owl pellets can tell scientists the numbers and kinds of small prey that live in an area.

The Barn Owls of Homestead Cave

Scientists found Homestead Cave while studying the Great Basin desert west of Salt Lake City, Utah. Homestead Cave is a cave where barn owls roost. Inside the cave scientists found piles of owl pellets. They studied the pellets for 3 years to find out what prey the owls had eaten.

The owl pellets in the cave were piled 2 meters (m) deep. The scientists dated the bones and discovered that the owl pellets had been piling up for the past 10,000 years! Owls that had lived thousands of years ago had roosted in the cave and spit up the oldest of the pellets. The oldest pellets were at the bottom layers of the pile. The darkness and cool temperature in the cave had preserved the pellets for thousands of years. No humans had disturbed this protected cave.

An owl pellet

Scientists identified the bones of 22 kinds of small mammals. They found wood rats, mice, voles, rabbits, and shrews of many kinds. Scientists figured out which animals lived in the area at different times. They compared the bones in each layer of the pile to animals living there now. They determined how the numbers and kinds of animals in the community changed over time.

Then scientists added climate data to their findings. They made a table of the climate and listed the animals that lived during that time. They observed how the numbers and kinds of animals changed as the climate changed over 10,000 years. They discovered that some kinds of small mammals were present only during times when the climate was warmer. When the climate was colder, different small mammals were common in the area.

Rebecca Terry was one of the biologists studying Homestead Cave. She was a student in college at the time. Her findings are helping biologists predict how climate change in the future will impact the small mammals in the Great Basin. This information will help scientists develop plans to protect the habitat of the animals living in the Great Basin.

Biologist Rebecca Terry studies owl pellets to reconstruct ancient environments in the Great Basin.

Fossils

How do we know what plants and animals looked like a very long time ago? We look at **fossils**. Fossils are the remains of plants and animals that lived a long time ago. Scientists study fossils to learn about past environments. Scientists who study fossils are **paleontologists**.

Dinosaurs lived a long, long time ago. No dinosaurs are living today. But scientists know a lot about dinosaurs by studying their fossils.

One of the most famous dinosaur fossils is named Sue. Sue is a *Tyrannosaurus rex*. When Sue died, about 66 million years ago, she might have fallen into a swamp. Soon after that, she was completely buried in mud. Over a period of thousands of years, her bones were replaced with minerals. As the mud around her turned into **sedimentary rock**, her bones became **petrified**, and turned into stone.

This is what Sue's bones looked like when she was found in South Dakota.

A scientist puts plaster over Sue's bones in preparation for transportation to the lab.

Scientists dug Sue out of the ground very carefully. In the lab, Sue's fossil bones and teeth were carefully cleaned. After a lot of hard work, all of Sue's bones and teeth were ready to put together.

Brushing sand away from a dinosaur claw fossil

A paleontologist working on a dinosaur fossil

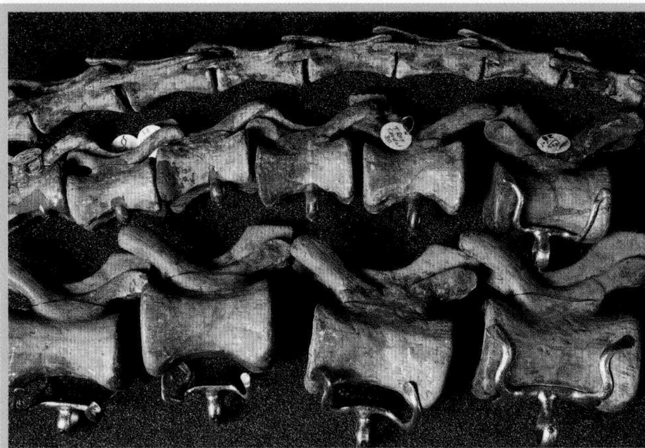

Tyrannosaurus rex **tail bones and teeth**

Sue's skeleton after it was assembled

It took a long time to get all the fossil bones in the right places. Finally, the bones all fit together. Everyone can now see what Sue's skeleton might have looked like.

After the skeleton was together, scientists wanted to know what Sue looked like when she was alive. They used pretend muscles, skin, and eyes on a copy of her skeleton to make her look real.

This is what Sue might have looked like.

Tyrannosaurus rex dinosaurs lived in forested river valleys in North America. Compare the environment shown in this photo to the current South Dakota environment where the fossil remains of Sue were found. Which one do you think is more like the past environment where Sue lived?

Dinosaurs have not lived on Earth for millions of years. But related animals that look like dinosaurs are found on Earth today.

Lizards, such as iguanas, look like dinosaurs. But when you look closer, you find many differences in their structures. Lizards have bent legs that sprawl to their sides. Dinosaurs, such as *Stegosaurus*, had straight legs and walked with their legs underneath them.

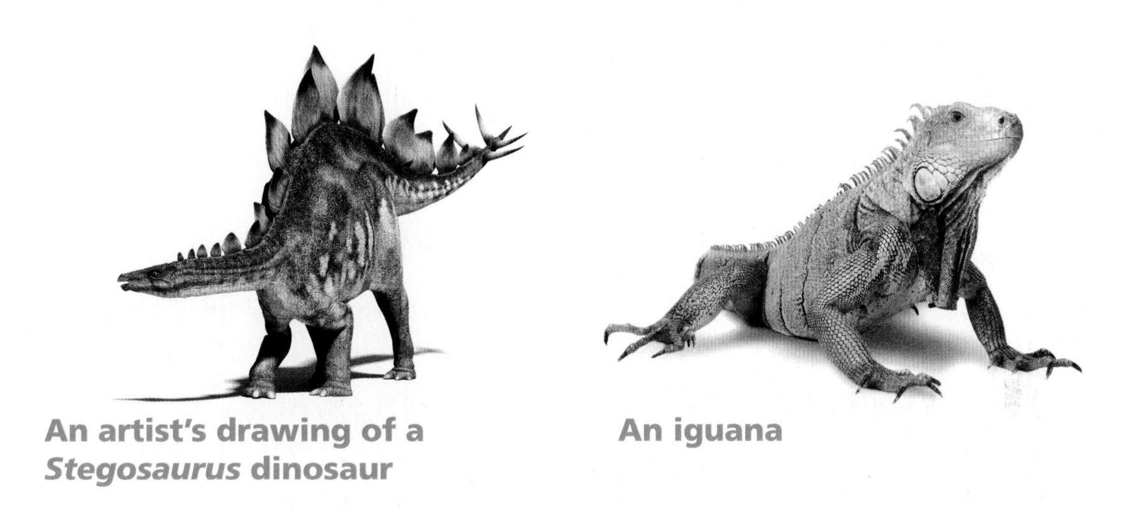

An artist's drawing of a *Stegosaurus* dinosaur

An iguana

The animals living today that are more closely related to dinosaurs are birds. Similarities between dinosaurs and birds, include hollow bones, a long S-shaped curved neck, an eggshell structure, and many skeletal structures. Birds are sometimes referred to as living dinosaurs. Modern birds, such as the emu, may have evolved from small meat-eating dinosaurs.

An emu

An artist's drawing of a bird-like dinosaur

Fish also lived when dinosaurs roamed Earth. Fossils of fish bones are found in rock. Fish fossils are found far from the ocean, lakes, and rivers.

Sometimes when fish died, they fell to the bottom of the ocean. When **sediments** covered the dead fish, the same thing happened to the fish that happened to Sue. Slowly, over thousands of years, the fish bones turned to stone as the sediment around the fish turned to rock.

Years later, strong earth forces lifted the ocean floor where the fish fossil formed. The fish fossil was lifted by these strong forces to a new location high in the mountains. That's why fossils of sea shells and fish can be found in the hills far from the ocean where they lived millions of years earlier.

Fossils provide us with **evidence** about the kinds of plants and animals that lived on Earth millions of years before there were people to see them. How do you think the environment was different when these fossil trees were alive 250 million years ago?

The Paleontologists' Puzzle

Paleontologists have to be good puzzle solvers. The environments in which they dig up fossils are often in arid, barren hills. They try to figure out what kind of environments the plants or animals they discover lived in when they were alive.

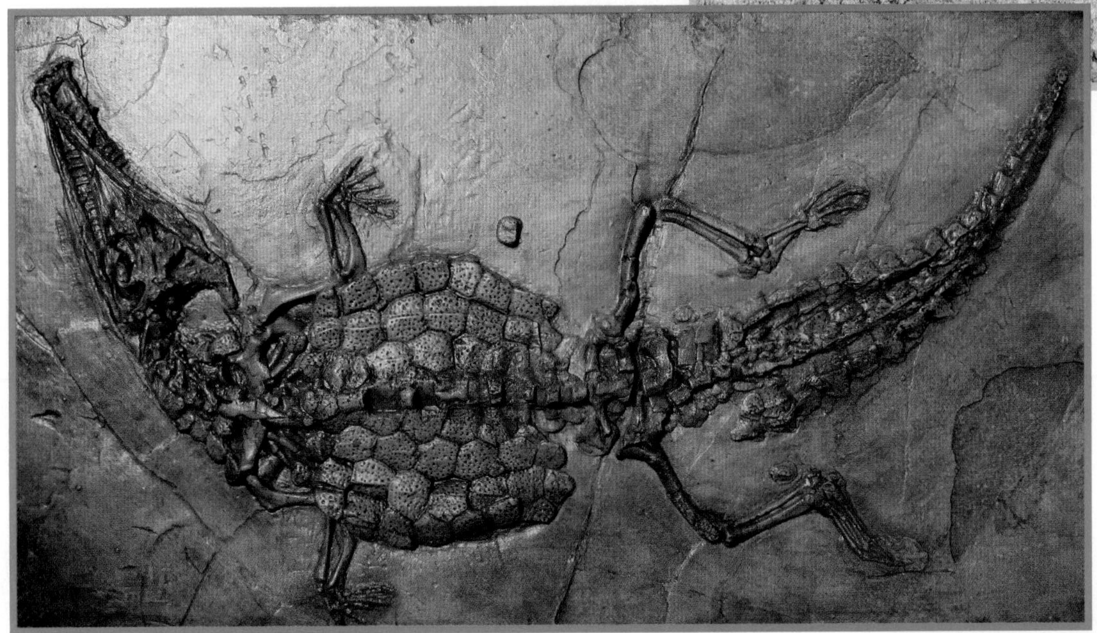

Thinking about Fossils

1. What are fossils?
2. Why are fossils of marine organisms, such as fish, sometimes found in arid, barren hills?
3. What can paleontologists learn from fossils?
4. Look at the images of fossils in the table on the next page. Try to determine the environment in which the organisms lived thousands or millions of years ago.

Fossil	What environment did it come from?
1. 	a. Ocean bay or shallow coast b. Mountain river or lake c. Grassland or meadow d. Forest pond or stream
2. 	a. Ocean bay or shallow coast b. Moist, shady woodland c. Hot rocky, sandy desert d. Tropical rain forest
3. 	a. Ocean bay or shallow coast b. Moist, shady woodland c. Hot rocky, sandy desert d. Tropical rain forest
4. 	a. Ocean bay or shallow coast b. Moist, shady woodland c. Hot rocky, sandy desert d. Tropical rain forest

Skeletons on the Outside

Not all animals have skeletons on the inside. A very large number of animals have skeletons on the outside of their bodies. This kind of skeleton is called an **exoskeleton**. Exoskeletons are made of hard, thin tubes and plates. All animals with exoskeletons are invertebrates, or animals without backbones.

Who Am I?

Match each animal with its description. Check your answers below.

1. Even the eyes of this animal are covered with a tough exoskeleton.

2. Like all arthropods, this aquatic animal must shed its exoskeleton to grow. It hides from its enemies while its new coat of armor hardens.

3. This land animal's exoskeleton has many different sections. Joints between the sections allow this animal to move easily.

4. This animal's exoskeleton creates a hard shell all around it. The two parts of the skeleton, called valves, are opened and closed by two big, strong muscles.

Scorpion

Tarantula

Clam

Crab

Animals are 1, tarantula; 2, crab; 3, scorpion; 4, clam.

Bony Comparison

Which is better, bones on the outside or bones on the inside? Look at the chart below to decide.

	Internal Skeleton	Exoskeleton
Protection	Protects inner organs. Does not offer protection from enemies.	Protects inner organs. May offer protection from enemies.
Growth	Grows and expands with age.	Does not grow. Sheds exoskeleton to grow.
Movement	Works with muscles and joints to allow for a variety of movements.	Plates and tubes of the exoskeleton are joined. Inner muscles provide movement.

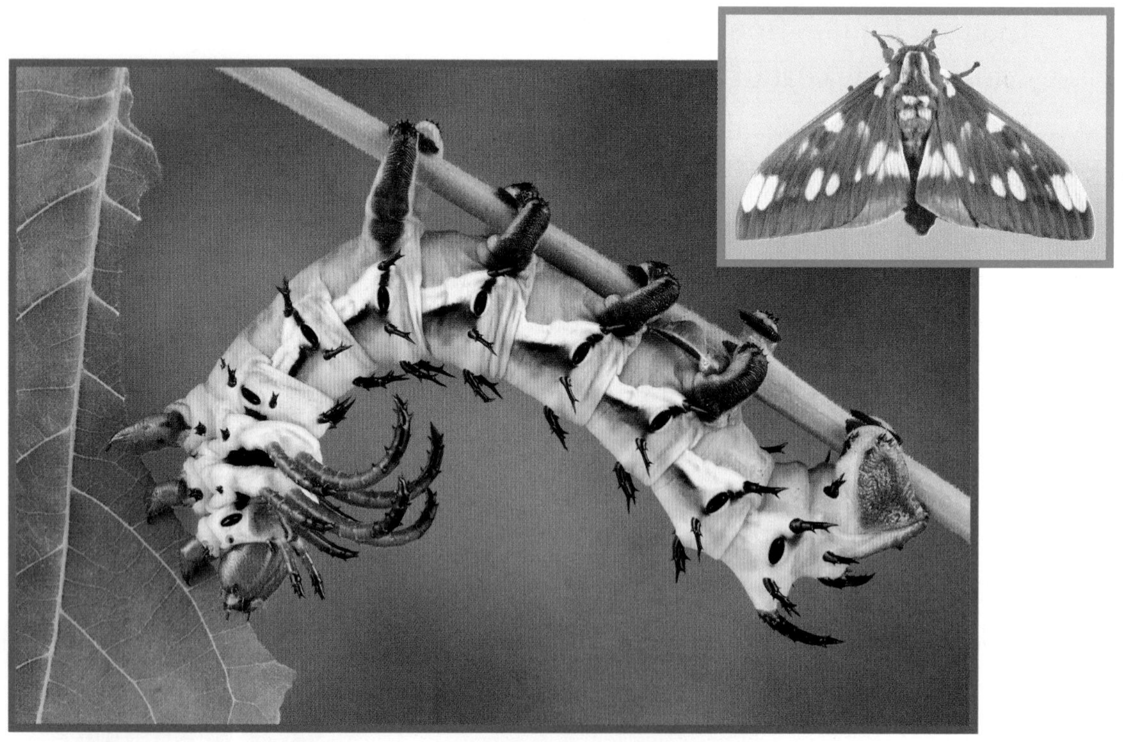

The hickory horned devil caterpillar is the larva of the regal moth. The caterpillar has an exoskeleton.

Crayfish, Snails, and Humans

There are many similarities between a crayfish, a snail, and you! There are many differences, too. Let's take a look at what is the same and what is different.

Skeleton

A crayfish's skeleton is on the outside of its body. It protects the crayfish from predators and other dangers. This exoskeleton doesn't grow with the crayfish. The crayfish has to shed its too-small shell to grow. A crayfish can also grow a new leg or claw if the old one is lost or damaged.

A snail has an outer shell that protects it from predators, too. A snail's shell grows along with the snail's body for the first 2 years of its life. The snail never sheds its shell. If a snail's shell cracks or breaks, it does not grow a new one.

Your skeleton is inside your body. It provides structure, gives your body shape, protects your internal organs, and allows for movement. Your skeleton keeps growing along with you. You can't grow a new arm or leg, but if you break a bone, new bone tissue will usually grow to heal the injury.

What structures are similar on crayfish, snails, and humans? What structures are different?

Internal Organs

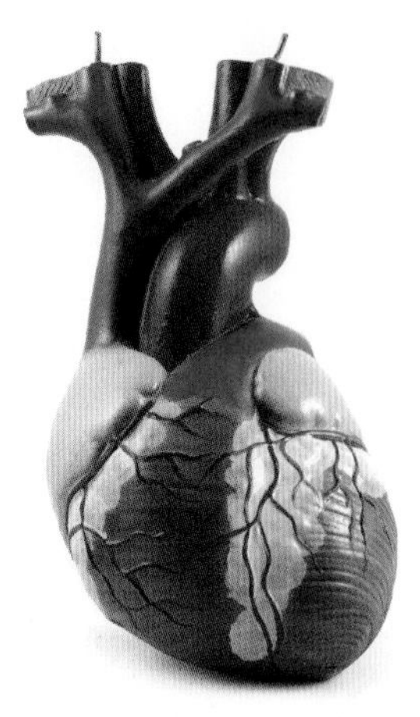

A model of a human heart

Crayfish, snails, and humans all have a heart to pump blood. They all have a stomach to digest food and organ systems to excrete wastes.

Humans and land snails have lungs to breathe air. Crayfish usually live in the water. Instead of using lungs to breathe, they take their oxygen out of the water through gills. The gills are tucked up under the carapace where the legs attach to the body.

Limbs

Crayfish have five pairs of legs. They can walk quickly in any direction on four of these pairs. This helps them move along the bottoms of ponds or streams as they look for food or avoid predators. The fifth pair of legs, located near the head, has large pincers. These pincers are more like arms than legs. The crayfish uses them to pick up food and defend itself.

Snails have no arms or legs. A snail moves with a muscular foot on the bottom of its body. This foot allows the snail to glide over almost any surface.

Humans have two legs to walk, run, and climb. They have two arms to pick up and carry things.

How do snails and humans move?

Your Amazing Opposable Thumbs

You've got two! They're amazing! They are your opposable thumbs. The thumb is the key to how humans hold things. No other living thing has a hand and thumb exactly like yours. But what is an opposable thumb?

An opposable thumb allows you to touch the tip of your thumb to the tip of each finger. Try it! Then try to touch the tip of your index finger to the tip of your pinkie. Your fingers are designed to work together with your thumb. This allows you to use your hands in many different ways.

Because we can hold things between the thumb and other fingers, we are able to pick up even tiny objects. This is called a precision grip. How important is your thumb? Try picking up a pen and writing without using your thumb.

What makes the thumb work the way it does? A unique joint connects the thumb to the palm. This joint is called a saddle joint. The thumb is the only place the saddle joint is found. This joint allows the thumb to move side to side and back and forth. The human saddle joint is very strong.

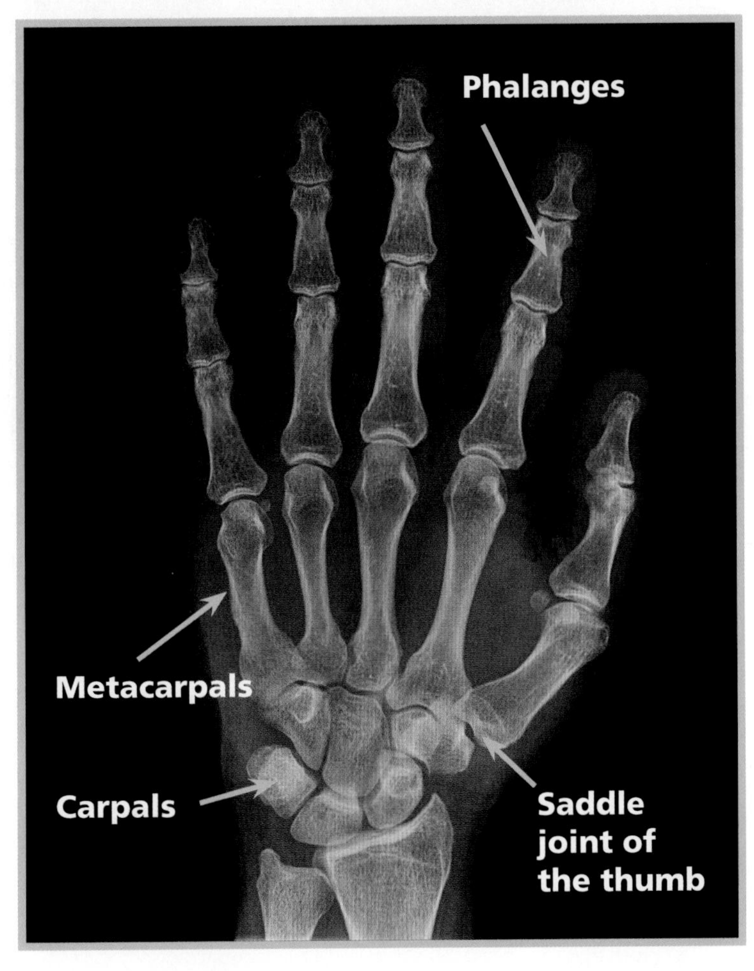

Joints and Muscles

Some parts of the body are quite flexible. Others move only a little. That's because there are different types of joints in different places. There are more than 200 joints in the human body. Each has its own job to do. The shape of a joint determines exactly how that part of the body will move. In general, the less movement a body part has, the stronger that part of the body is.

A slippery, smooth tissue covers the ends of the bones where they meet and touch. This tissue is cartilage. The cartilage, kept slippery by a special fluid, allows the bones to move against one another with less rubbing.

Hinge Joints

Hinge joints are simple but important joints in your body. A hinge joint works like the hinge of a door. It allows movement in only one direction. Hinge joints allow your legs, arms, and fingers to bend and straighten.

Hinge joints are found at the knees, elbows, fingers, and toes. The knee joint is the largest hinge joint in the human body. It "locks" when you stand straight. The knee joint's locking action makes it easier for you to stand for long periods of time.

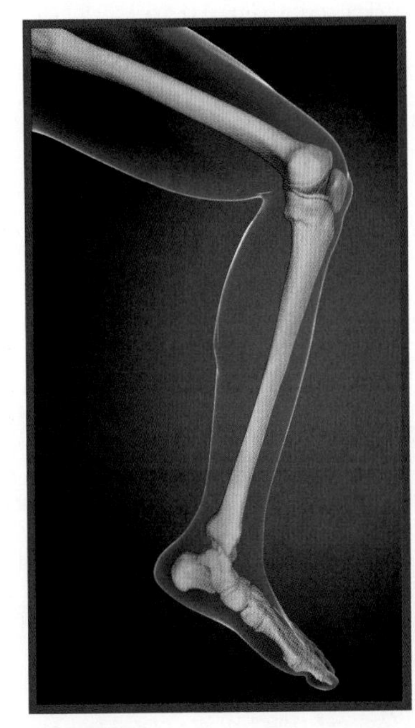

Hinge joints allow your knees to bend.

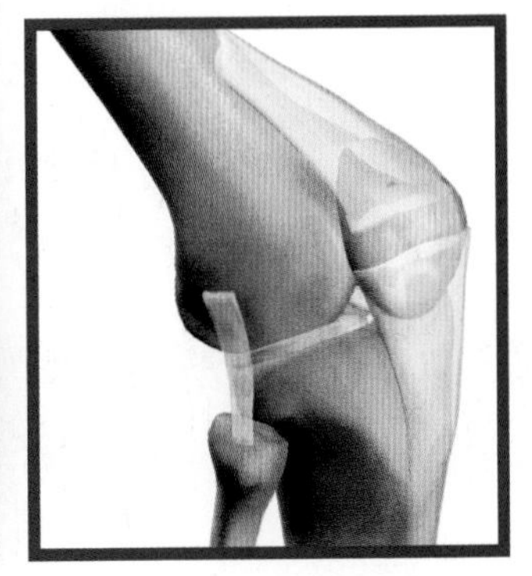

Ball-and-Socket Joints

Ball-and-socket joints allow bones to swivel in nearly any direction. Ball-and-socket joints get their name from the shapes of the two bones that meet there. One example is the shoulder joint. The ball of the upper arm bone fits snugly into the hollow socket of the shoulder.

Ball-and-socket joints are found at the shoulder and hip. The hip is the strongest of all joints. It must be strong to support the weight of the upper body. It is not quite as flexible as the shoulder joint. You can swing your arm in a complete circle.

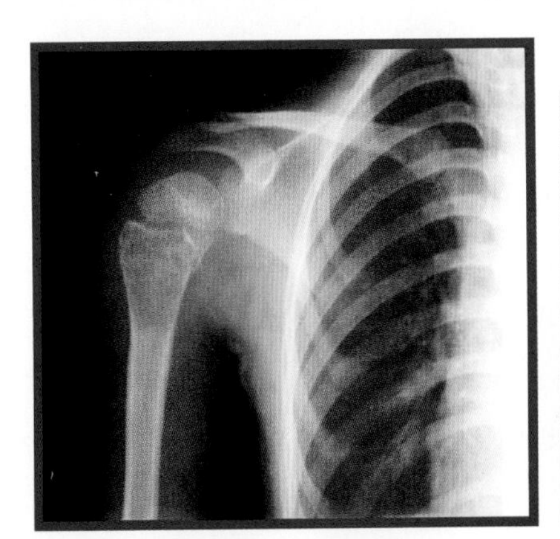

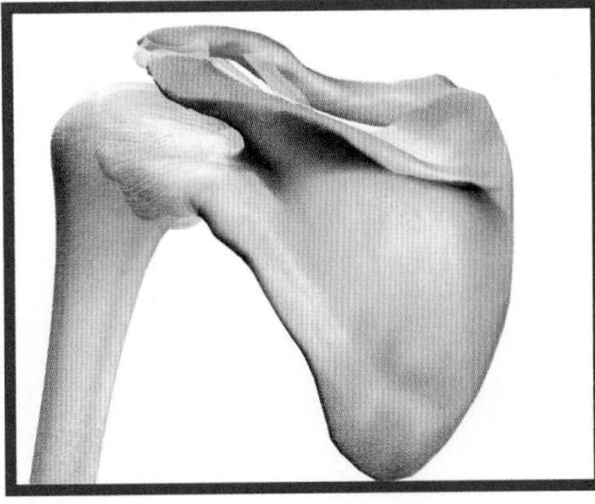

Ball-and-socket joints allow you to swivel your shoulders and hips.

Gliding Joints

Gliding joints have two flat surfaces that glide smoothly and easily past one another. These joints allow only small movements. Gliding joints are found in the neck and spine, between pairs of vertebrae. Other gliding joints are found in the wrists and ankles.

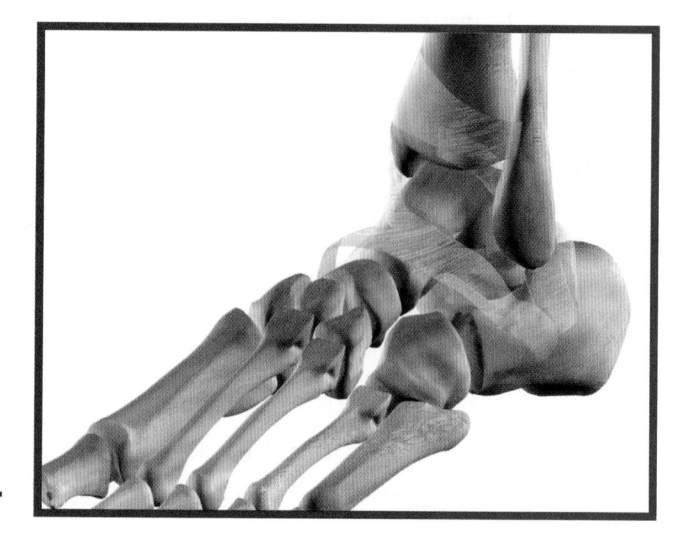

Gliding joints allow your ankles to move.

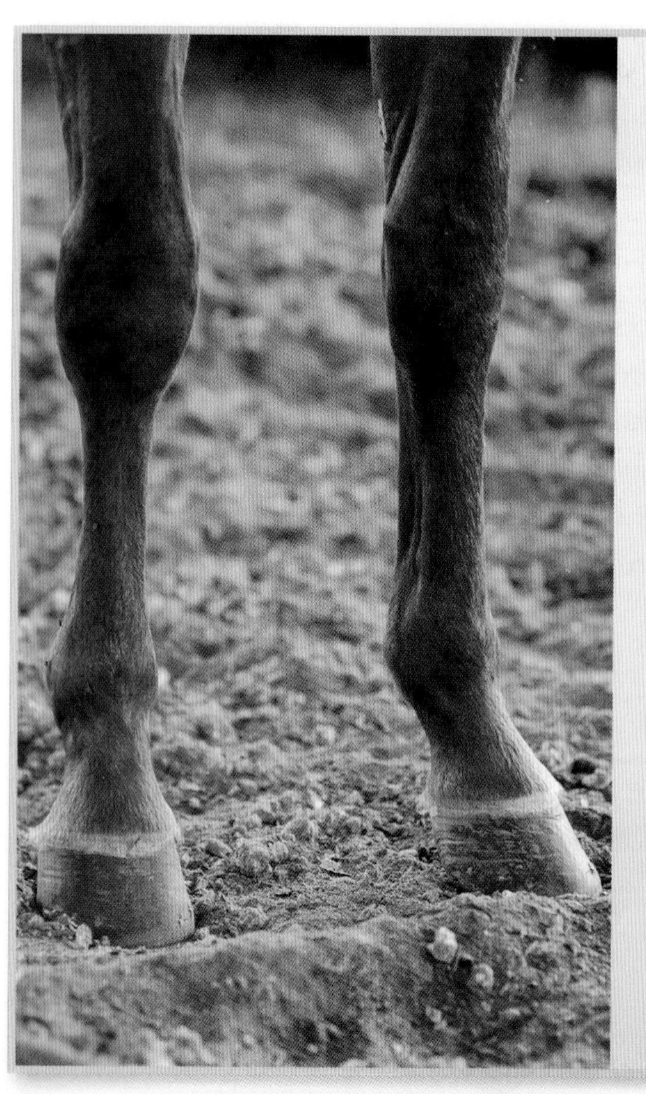

Interesting Animal Joints

How do other animals' joints compare to human joints?

Horse Horses have special tendons that work with the carpal joints in their front legs to prevent bending. This helps the horse to stand, and even sleep standing for hours.

Gibbon In addition to having opposable thumbs, gibbons also have an opposable toe on each foot. This toe works like the opposable thumb. It allows the animal to hang onto branches with its feet while it travels from tree to tree.

Goat Goats and some other animals that eat plants have unusual jaw joints. These jaw joints allow a goat to move their jaws sideways, up and down, and front to back when chewing.

Muscles

Muscles help you run and kick soccer balls.

Your body has more than 700 muscles. Without these muscles, you'd be going nowhere! Every move you make is powered by muscles. Muscles help you walk, run, and hit a baseball. When you blink, chew, or talk, you are using your muscles. Muscles also help keep your body upright and make your movements steady.

Muscles are made up of small, thick bundles of fibers. These fibers are designed for movement. When muscles contract, they pull the bones, causing movement.

About 650 of your muscles are skeletal muscles. Skeletal muscles move the arms, legs, and other parts of the body. Skeletal muscles are also called voluntary muscles. That's because you can control these muscles. There are two other types of muscles. They are smooth muscles and cardiac muscles. Smooth muscles are found in the walls of blood vessels and some organs. Cardiac muscles are found in the walls of the heart.

Muscle Pairs

Skeletal muscles nearly always work in pairs or groups. While one muscle contracts, the other relaxes. Look at your upper arm as you bend your arm at the elbow. The biceps and triceps muscles in your upper arm are working together. The biceps contracts and becomes shorter, while the triceps relaxes and becomes longer.

triceps
biceps

It's important to take care of your muscles. This means getting plenty of exercise and eating well. The more your muscles are used, the stronger they will become and the better they will work.

Muscles on the Move

Facial muscles There are about 30 different muscles in your face. Most facial muscles are attached to each other or to the skin, not to bone. The facial muscles control a variety of movements. When you raise your eyebrows, wrinkle your forehead, close your eyes, or smile, your facial muscles are at work.

Neck muscles Muscles in the neck must be very strong. They have to keep the head upright. An adult human head weighs about 4.5 kilograms (kg).

Hand muscles Each hand has about 20 different muscles. With so many muscles, the hand can move in a variety of ways.

Abdominal muscles Your abdominal muscles allow you to twist and bend your body. They also help you inhale and exhale.

Gluteus maximus The gluteus maximus is the largest muscle in your body. It is also one of the strongest. This big muscle helps you run, jump, and climb. It's also the muscle on which you sit!

Leg muscles The muscles in your thigh bend, straighten, and twist both your hip and your knee. The muscles in your calf allow you to bend, straighten, and twist your ankle.

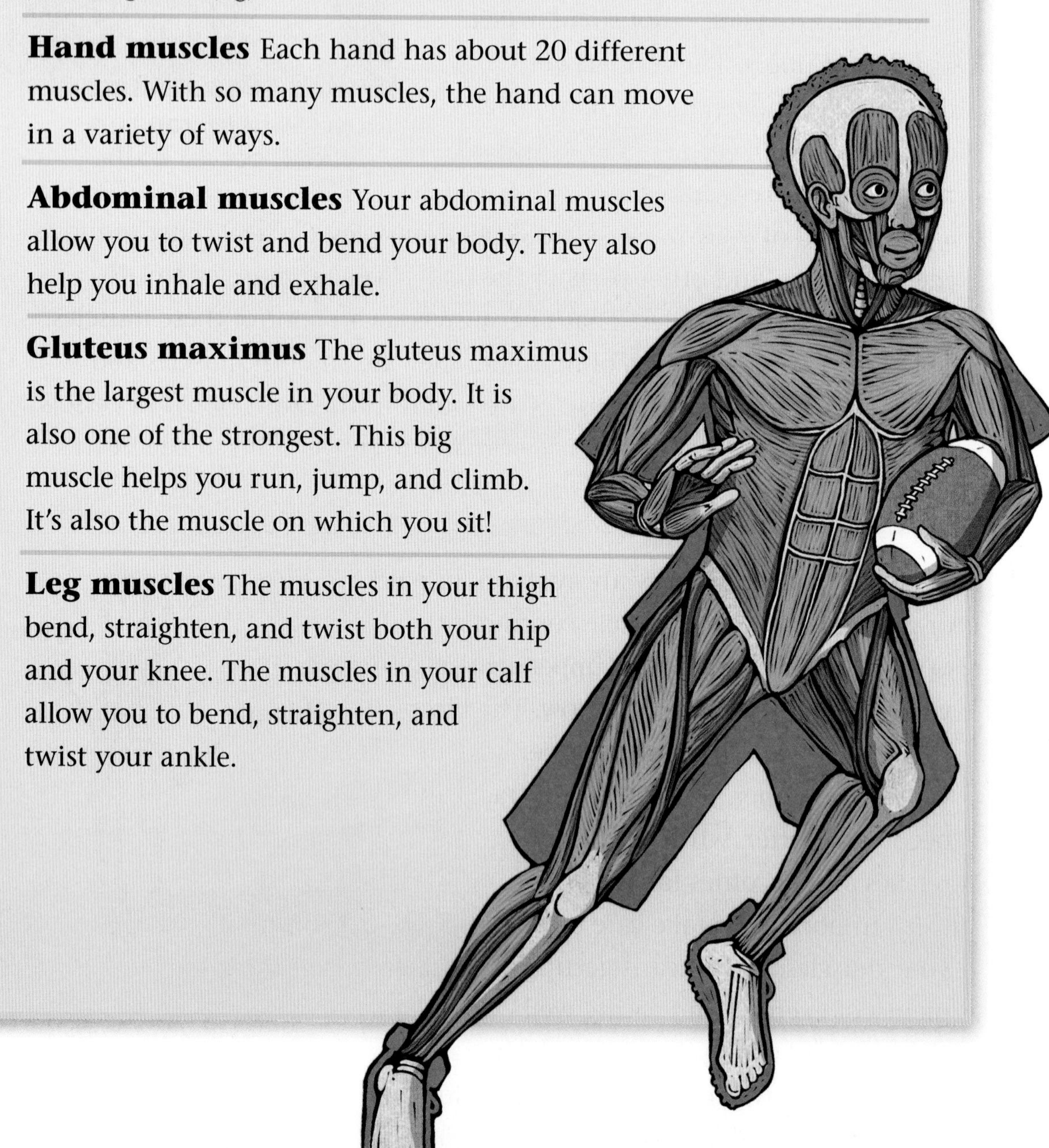

Muscles and Bones Working Together

Check out the muscles and bones that work together to make movement possible.

Gluteus Maximus—Thighbone and Pelvis The gluteus maximus connects the femur, or thighbone, to the pelvic bones. This muscle controls running, jumping, and climbing.

Biceps and Triceps—Arm Bones The biceps and triceps muscles help move the arm bones. The arm bones are the humerus, radius, and ulna. Bend your arm at the elbow, and the biceps contracts while the triceps relaxes. Straighten your arm, and the opposite happens.

Various Muscles—Shoulder Blade and Upper Arm Bone The shoulder is one of the most flexible parts of the body. Many muscles are needed to hold the scapula, or shoulder blade, in place. The deltoid is one of the larger muscles in the shoulder area. This muscle helps raise the arm.

Calf Muscles—Heel Bone and Lower Leg Bones The heel bone is connected to the calf muscles by the Achilles tendon. It is the longest and strongest tendon in your body. Calf and shin muscles also connect your lower leg bones, the tibia and fibula, to your ankle. These muscles help bend and straighten the ankle.

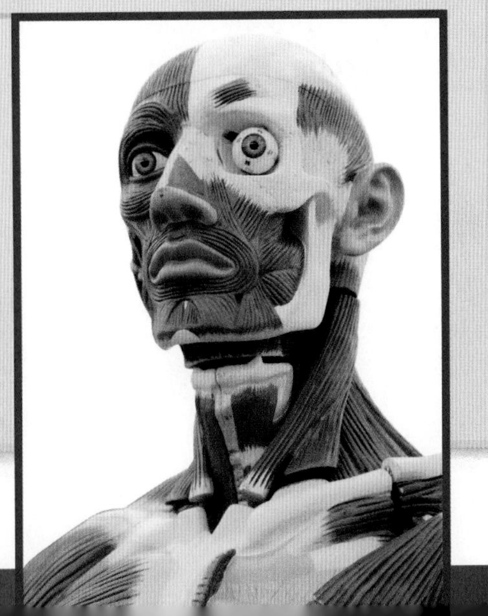

Neck Muscles—Skull and Spine Pairs of muscles in your neck connect your skull to your spine. Each pair of muscles moves the skull in a different direction.

Fingerprints

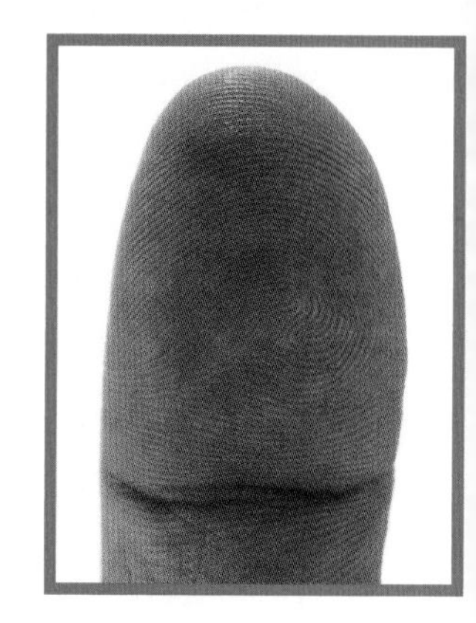

Take a close look at the tips of your fingers. Can you see swirling lines? These lines are made by ridges and furrows in your skin. The ridges help you grip objects. Without them, things might slip from your hands. Everyone has these ridges. They are your **fingerprints**. Your fingerprints grow larger as your body grows, but they do not change in any other way. Your fingerprints are unique. No one else has exactly the same design. Not even identical twins have fingerprints that are exactly the same. The Chinese started using fingerprints as marks of identification around 200 BCE.

The shape of a fingerprint is called its pattern. Fingerprints can be separated into three general patterns. They are arches, loops, and whorls. Ridges in an arch start on one side of a finger, rise and fall in the center of the finger, then end on the other side of the finger. Ridges in a loop also start on one side of the finger. But they rise and curve back to end on the same side they started. A whorl is a set of circles inside each other, formed by ridges.

Arches

Loops

Whorls

What Type of Fingerprints Do You Have?

- About 5 percent of all fingerprints are arches.
- About 30 percent are whorls.
- About 65 percent are loops.

Many people share the same fingerprint pattern, but details in the pattern make one fingerprint different from another. Scientists who study fingerprints look at ridge endings, ridge fragments, and places where ridges split. They notice how these details are positioned. This is what makes each fingerprint unique.

In the 1880s, Sir Francis Galton (1822–1911) observed that each person's fingerprints are different. He claimed they would not change. That was the beginning of fingerprint science. Because fingerprints are unique, they can identify people. The fingerprints of crime suspects can be compared to prints left at the scene of a crime. The first case known to be solved by fingerprints was in 1892. There was a murder in La Plata, Argentina. Fingerprints at the scene belonged to a woman in the house. She had accused a neighbor of the crime. Faced with the fingerprint evidence, the woman confessed.

In 1897, Sir Edward Richard Henry (1850–1931) set up a system for classifying fingerprints. Henry was London's assistant commissioner of police. The Henry system compares inked fingerprint cards on file. It identifies people through their fingerprint patterns. The Henry system is still used today.

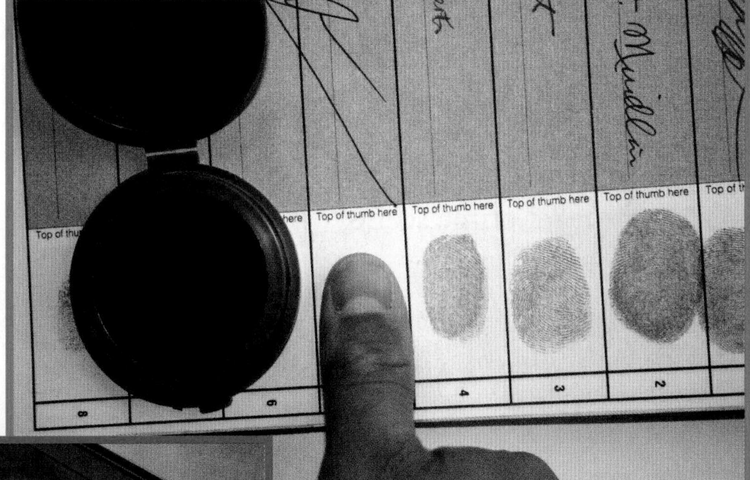

Making an inked print

Making a digital print

A police officer looking at a fingerprint on a computer

The Federal Bureau of Investigation (FBI) started to compile fingerprint files in 1924. There are millions of fingerprints in the FBI fingerprint files. Fingerprints are still sometimes made by rolling fingers in ink and pressing them against paper. New techniques for making digital images of fingerprints are being developed. FBI fingerprints are divided into criminal and noncriminal files. Noncriminal prints include government employees, teachers in some states, and people who have volunteered their prints for identification purposes.

Latent prints are fingerprints we leave on certain surfaces. Latent prints are used to connect criminals to their crimes. Criminals do not leave prints at crime scenes on purpose. In fact, their fingerprints are usually invisible. They're made by oil or sweat on the skin ridges.

Footprints

Fingertips are not the only places on the skin with ridges. Ridges also exist on your palms, your toes, and the soles of your feet. These ridges are also unique to you. They can be used to identify you. That's why hospitals take inked prints of the soles of babies' feet.

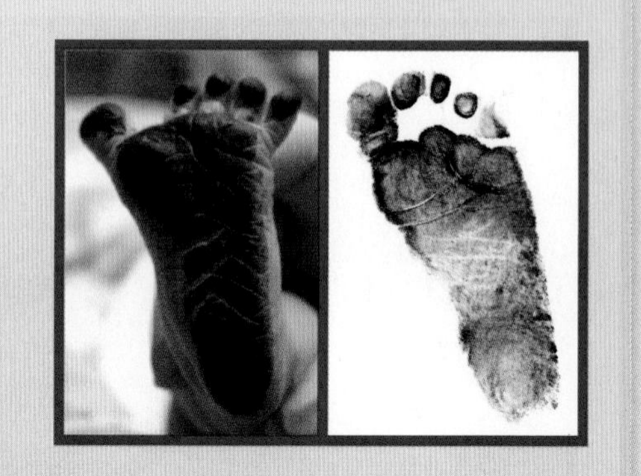

Dusting for fingerprints

Forensic scientists have ways to make latent prints visible. Prints might appear when they are brushed with special powders. The powders stick to oil and sweat. Ninhydrin is one of many chemicals used to see latent prints. It works by reacting with acids in sweat. Prints can then be photographed and matched with prints on file. In that way, criminals can be identified.

Forensic science has been helped by tools such as the automated fingerprint identification system (AFIS). Using computers, AFIS can compare even a small portion of a print against millions of fingerprints on file. AFIS uses ridge characteristics to make a list of matches. The final comparisons are made by fingerprint scientists.

Thinking about Fingerprints

1. What are the three most common fingerprint patterns?
2. What is the difference between loop and arch fingerprint patterns?
3. What is the difference between a latent print and an inked print?

DNA

All living cells contain a genetic material called **DNA** (deoxyribonucleic acid). Your DNA is responsible for who you are and how you look.

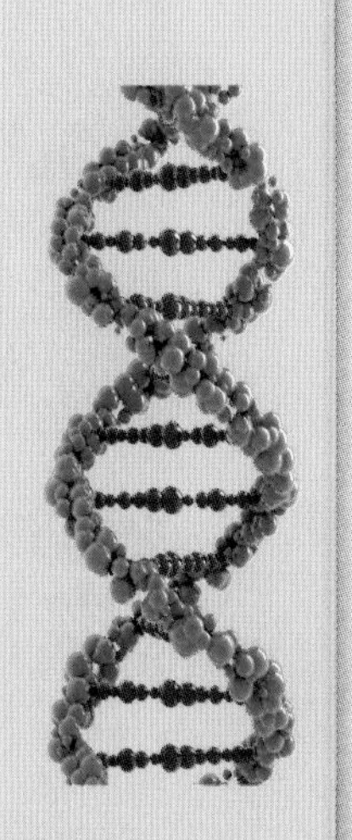

Everyone's DNA is made of the same four chemical units called nucleobases. But those chemical units are arranged in a different sequence in each person's DNA. That means DNA can be used to identify people the same way fingerprints can be used.

DNA sequencing analyzes and compares DNA from different sources. It's used to determine which people are related. It can also connect a person to DNA left at the scene of a crime.

Supertwins

Fingerprint experts tell us that every person's fingerprints are unique. That means no other person in the world has fingerprints like yours. Not your mother or father. Not your brothers or sisters or any other relatives. But is that really true?

Think about identical twins. Identical twins are more than brothers or sisters who were born at the same time. Identical twins have exactly the same DNA. This means they are actually two copies of the same person. Identical twins have the same color eyes and the same color hair. They walk the same, talk the same, and have hands that are the same shape.

So what about identical twins' fingerprints? If they really are two copies of one person, they might have the same fingerprints. You can find out for yourself. Meet the Ferreira supertwins.

The Ferreira brothers, Matt, Jeff, and Dan, are triplets. They are identical triplets. Matt, Jeff, and Dan are three copies of the same person. Another name for more than two copies of the same person is supertwins.

Matt, Jeff, and Dan Ferreira

Here are the fingerprints from the Ferreira family's left hands. What can you figure out? Are they the same in any way? Are they different in any way?

Let's see how good you are at fingerprint detective work. Each set of prints in the boxes below belongs to Matt, Jeff, Dan, or their dad. But the prints aren't labeled with the name of the person who made them. Which prints do you think belong to the supertwins? Which prints belong to their dad? After you make your predictions, check your answers.

Do identical twins or supertwins have identical fingerprints? Now you know.

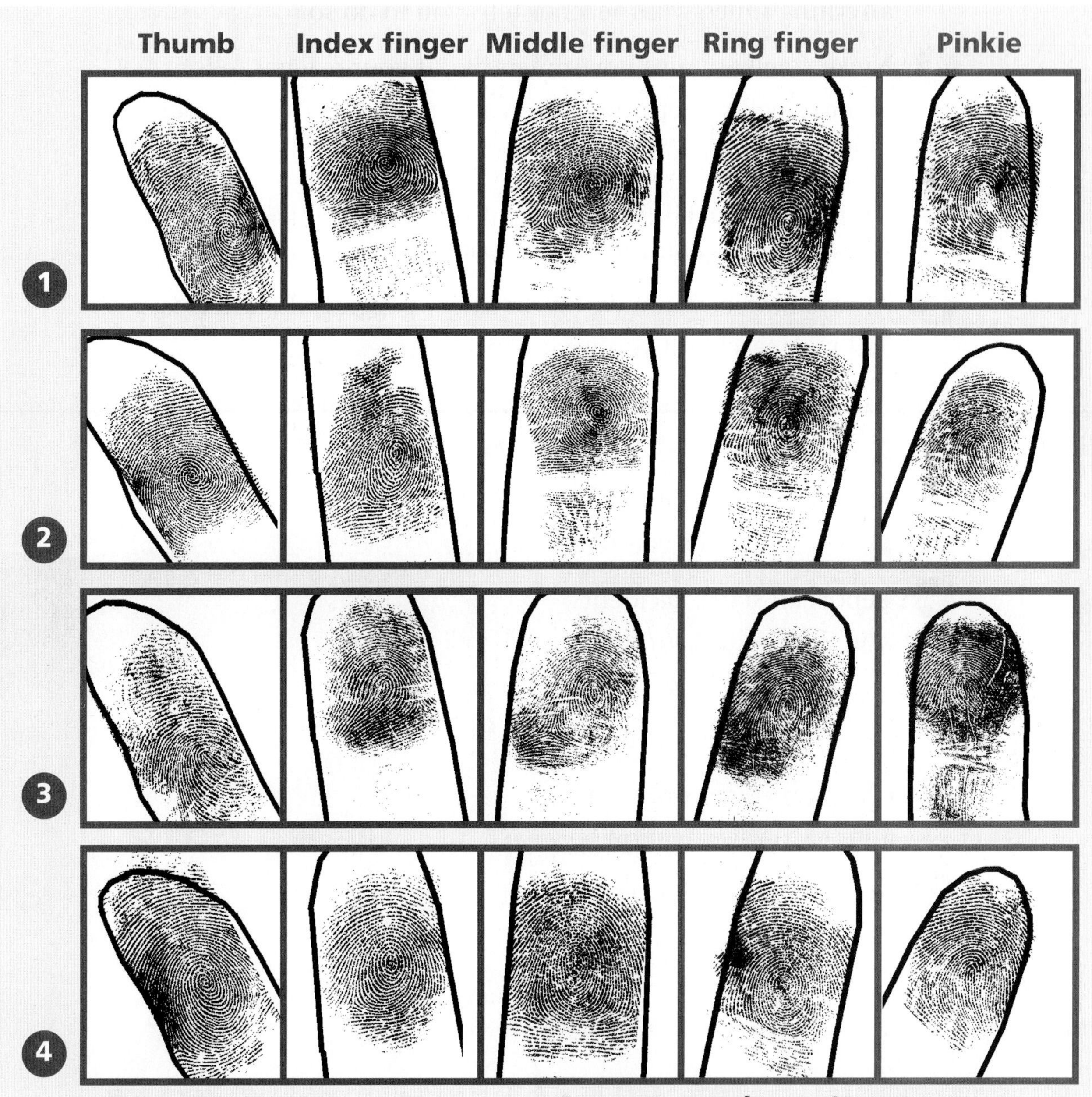

Print sets 1, 2, and 4 belong to the supertwins. Set 3 belongs to their dad.

Science Safety Rules

1. Listen carefully to your teacher's instructions. Follow all directions. Ask questions if you don't know what to do.
2. Tell your teacher if you have any allergies.
3. Never put any materials in your mouth. Do not taste anything unless your teacher tells you to do so.
4. Never smell any unknown material. If your teacher tells you to smell something, wave your hand over the material to bring the smell toward your nose.
5. Do not touch your face, mouth, ears, eyes, or nose while working with chemicals, plants, or animals.
6. Always protect your eyes. Wear safety goggles when necessary. Tell your teacher if you wear contact lenses.
7. Always wash your hands with soap and warm water after handling chemicals, plants, or animals.
8. Never mix any chemicals unless your teacher tells you to do so.
9. Report all spills, accidents, and injuries to your teacher.
10. Treat animals with respect, caution, and consideration.
11. Clean up your work space after each investigation.
12. Act responsibly during all science activities.

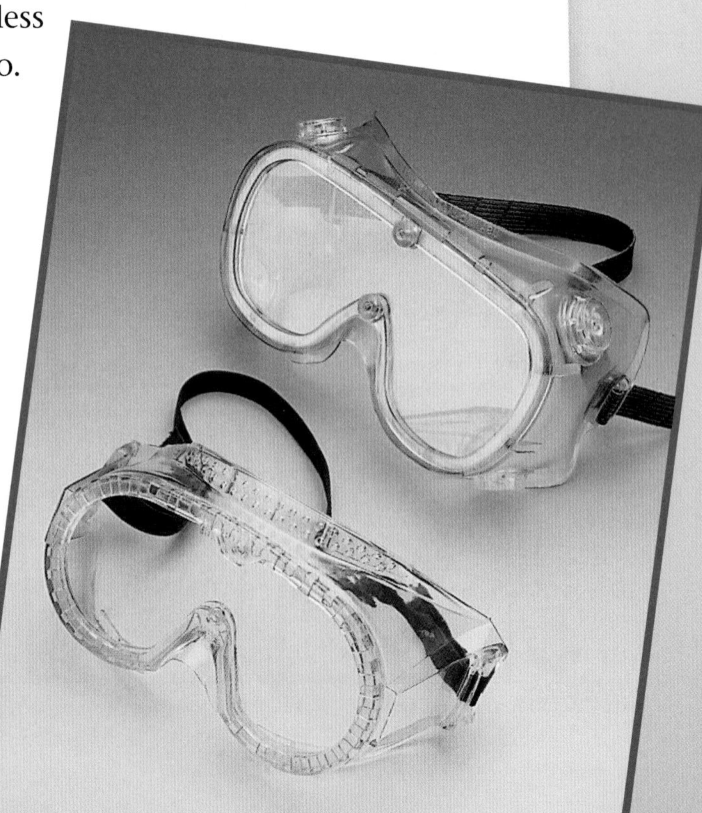

Glossary

adaptation any structure or behavior of an organism that allows it to survive in its environment

antenna (plural **antennae)** the thin feeler on the head of an animal like a crayfish, an isopod, or an insect

aquatic referring to water

behavior the actions of an animal in response to its environment

beneficial good or advantageous

biologist a scientist who studies living organisms

camouflage an adaptation that allows an organism to blend into its environment

carapace a hard outer shell that covers the main part of the body of an animal

carnivore an animal that eats only animals

cartilage the smooth, flexible material that connects some bones and gives shape to some body parts

chromosome a structure that carries genes

chrysalis the hard-shelled pupa of a moth or butterfly

contract to become smaller or shorter in length

cotyledon the plant structure that provides the germinated seed with food

crustacean a class of mostly aquatic animals with hard, flexible shells

detrimental harmful or bad

DNA (deoxyribonucleic acid) a material that carries the genetic messages of heredity

dormant inactive or resting

egg the first stage in an animal's life cycle

embryo the undeveloped plant within a seed

endanger to be at risk of becoming extinct

environment everything that surrounds and influences an organism

evidence data used to support claims. Evidence is based on observations and scientific data.

exoskeleton any hard outer covering that protects or supports the body of an animal

fingerprint the ridges in your skin at the tip of your fingers. Arches, loops, and whorls are fingerprint patterns.

flower a plant structure that grows into fruit

food chain a description of the feeding relationships between all the organisms in an environment

fossil any remains, trace, or imprint of animal or plant life preserved in Earth's crust

fruit a structure of a plant in which seeds form

function an action that helps a plant or an animal survive

gastropod the family of snails

gene a message carried by a chromosome

generation a group of organisms born and living at the same time

genetics the study of how living things pass traits to their offspring

herbivore an animal that eats only plants or algae

hibernate when animals sleep through the winter

inherited trait a characteristic that is passed down from generation to generation

invasive an organism that thrives in a new area but causes problems to the organisms in that ecosystem

joint a place where two bones come together

leaf **(**plural **leaves)** a plant structure that is usually green and makes food from sunlight, water, and carbon dioxide

life cycle the sequence of changes or stages an organism goes through as it grows and develops

ligament tissue that connects bone to bone

mast year a year when trees produce a lot of seeds

mature fully developed

migrate when animals move from places with cold weather to places with warm weather

molt to shed an outer shell in order to grow

muscle tissue that can contract and produce movement

nutrient a material needed by a living organism to help it grow and develop

offspring a new plant or animal produced by a parent

omnivore an animal that eats both animals and plants

organism any living thing

paleontologist a scientist who studies fossils

parent an organism that has produced offspring

petrify to change into stone over a long period of time

pincer an animal's claw used for grasping

population all organisms of one kind that are living together

predator an animal that hunts and catches other animals for food

prey an animal eaten by another animal

proboscis a long, strawlike mouth

protect to keep safe

pupa the stage of an insect's life cycle between the larva and the adult stages

reproduce to have offspring

riparian along a river or stream

root the part of a plant that grows underground and brings water and nutrients into the plant

sediment pieces of weathered rock such as sand, deposited by wind, water, and ice

sedimentary rock a rock that forms when layers of sediments get stuck together

seed the structure in a fruit that holds the undeveloped plant, or embryo

stem any stalk supporting leaves, flowers, or fruit

structure any identifiable part of an organism

survive to stay alive

swimmeret a small, soft leg under the tail of a crayfish

tendon ropelike tissue that connects muscle to bone

terrestrial referring to land

thrive to grow fast and stay healthy

Index